# EXPERIMENTAL STRUCTURAL DYNAMICS

## *An Introduction to Experimental Methods of Characterizing Vibrating Structures*

ROBERT E. COLEMAN

*1663 Liberty Drive, Suite 200*
*Bloomington, Indiana 47403*
*(800) 839-8640*
*www.authorhouse.com*

*First published by AuthorHouse 04/25/04*

*ISBN: 1-4184-1137-X (e)*
*ISBN: 1-4184-1138-8 (sc)*
*ISBN: 1-4184-1139-6 (dj)*

*Library of Congress Control Number: 2004106070*

*Printed in the United States of America*
*Bloomington, Indiana*

*This book is printed on acid-free paper.*

# Preface

It is not uncommon to find engineers in test labs or design groups who have not had occasion to use the mathematical tools acquired in college. When suddenly faced with vibration issues they find themselves ill equipped to get a solid grasp of the vibration process. It is the intent of this technical reference to provide access to vibration theory, initially at a very elementary level, progressing from basic analytical formulations toward the more mature mathematical representations associated with eigenvectors and the Fourier Transform.

Mode shapes are introduced without any reference to the eigenvalue problem, but connected immediately to simple coordinate transformations in two and three dimensions. This leads to a rather simple picture of operators, ultimately yielding a straight forward derivation of the Frequency Response Function (FRF) formula.

It is hoped that many engineers will find their way back into a more analytical approach to vibration problems. Experienced analysts and academics were given consideration as well, providing fresh viewpoints from time to time, such as the development of modal force as a contravariant vector, providing a detailed view of the FRF as a superposition of modal FRFs.

# Table of Contents

# Chapter I

# BASIC VIBRATION CONCEPTS

## 1.1 Introduction

This text is about vibrating structures. The structures considered could be any of a broad range of engineered products, from TV sets, computers and other electronics products to cars, trucks, trains, aircraft, rockets and other vehicles. We could be talking about bridges or buildings. All of these products have the potential to fail in their product performance without proper engineering to avoid damage that could be caused by mechanical vibration. Aircraft are analyzed and tested to arrive at structural design characteristics that are successful in handling the aerodynamic loads encountered during flight. Vehicles are subject to vibration and noise originating from the engine, tires rolling over an irregular surface at high speed and turbulent air flow over the body. Vibration design characteristics are designed into the vehicle to avoid wear and fatigue failure of certain components and to provide a comfortable and quiet ride for the passenger.

For general arbitrary structures, the vibration process is very complicated, so complicated that one might expect the process impossible to comprehend. Impossible, except for the ability to analyze the most complex vibration motion as a superposition of relatively simple processes. It turns out that no matter how complicated the structure and no matter how complicated the vibratory motion of the many parts of the vibrating structure, it is usually possible to separate the process into easily understood fundamental vibratory processes.

It is the goal of this text to first present the theory underlying the simple vibratory process, then develop the concepts allowing application of this understanding to the analysis of any complicated vibratory process for the most complex structure. There is one limitation in the level of structural complexity to be considered, however: The text will be concerned with linear structures. Vibration displacements will be small and stiffness characteristics will be fixed, independent of the amount of structural deformation.

## 1.2 Simple Harmonic Motion

A natural starting point is to study the motion of the simplest of structures in a natural state of vibration. Figure 1-1 depicts such a structure and the simple vibratory motion that results when a lumped mass sitting on a spring is made to vibrate freely. The mass is initially displaced upward from its equilibrium position on the spring. From this position it is released, accelerating downward under the pull of the stretched spring. The continuous motion of the mass is graphed with the solid curve in the figure. Instantaneous positions of the mass at key points in time are sketched. The mass is seen to oscillate, moving down until the upward force of the compressed spring brings the downward motion to a stop. Then the upward push of the compressed spring propels the mass upward until the cycle of oscillation is complete when the upward motion is stopped under the downward pull of the stretched spring. The cycle of motion is completed in one second in our example. From this point in time the mass will continue to oscillate in this fashion forever in the absence of any other influences, i.e, friction, human intervention, etc.

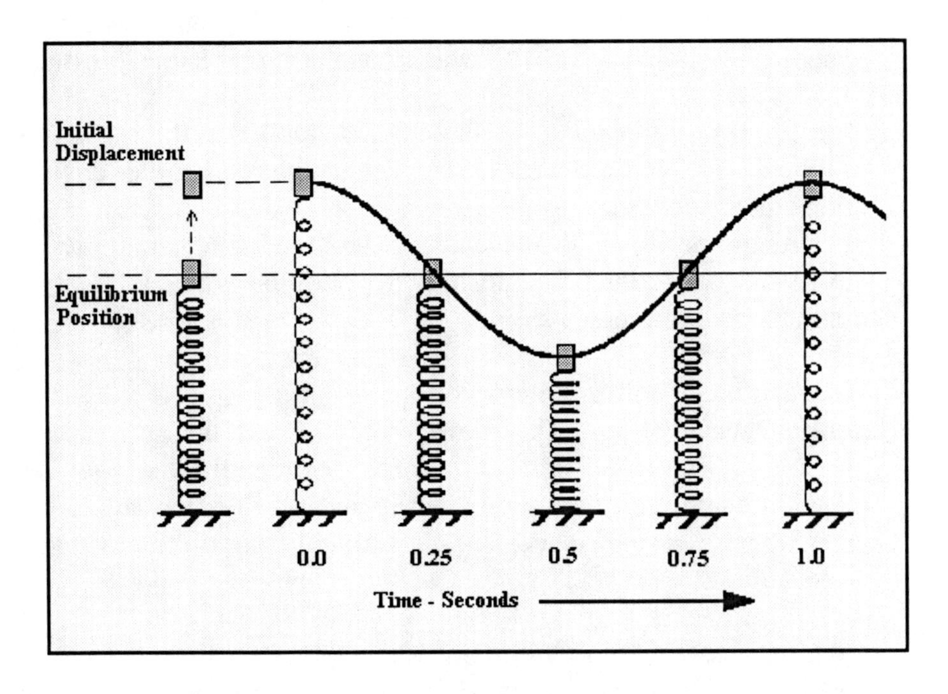


**Figure 1-1. The vibration motion of a mass on a spring. After being displaced upward from the rest position and released, the mass oscillates in simple harmonic motion. The time period of oscillation is one second in our example.**

Let's analyze the motion of the mass at each of the key points in time shown in Figure 1-1. The mass is released at the start of the process, 0.0 seconds. At the instant it is released, the stretched spring is exerting a pull causing an acceleration downward. However, the velocity is zero, the mass is at rest initially. So, at zero time we have the maximum positive displacement, zero velocity and maximum negative (downward) acceleration.

At 0.25 seconds the mass has returned to the equilibrium position, zero displacement. But now it is moving downward with maximum velocity. At the equilibrium position there is no force acting on the mass, so its acceleration is zero. Thus, the momentum of the mass will allow it to move right through the equilibrium position, after which its motion begins to be resisted as the spring is compressed.

The downward movement of the mass is stopped at 0.5 seconds after the continued compression of the spring overcomes the downward momentum, decelerating the motion to zero velocity. Now we have maximum negative displacement, zero velocity and maximum positive acceleration. The positive acceleration is due to the upward force of the compressed spring.

The continued upward thrust of the recoiling spring maintains an upward acceleration. The thrust diminishes as the spring uncoils, until at 0.75 seconds the spring has returned to the equilibrium position again, and the upward thrust and acceleration are zero. So now we have zero displacement, maximum positive velocity and zero acceleration.

Finally, one cycle of vibration is complete at 1.0 second as the upward momentum carries the mass upward to the maximum displacement position where it is momentarily brought to rest again by the downward pull of the stretched spring. The final conditions are the same as at the start of the process, when the mass was first released: Maximum positive displacement, zero velocity and maximum downward (negative) acceleration.

Figure 1-1 illustrates one way of initiating free vibration in a simple mass-spring system. The mass is given an initial displacement and released. Now, consider a way of initiating vibration without displacing the mass initially. Figure 1-2 depicts the initiation of vibration by striking the mass with a hammer, almost instantaneously imparting an initial maximum upward velocity to the mass, which is hanging on a spring, initially at the equilibrium position. For all practical purposes we can consider the time duration that the hammer is in contact with the mass to be infinitessimally small. The measure of smallness is relative to the time period of one cycle of oscillation.

The following conditions summarize the motion of Figure 1-2. At 0.0 seconds (initial condition) the displacement is zero, velocity is a positive maximum and acceleration is zero (after an infinitessimal time increment during which the impact transient accelerates the mass to the initial velocity). At 0.25 seconds displacement is at the positive peak, velocity is zero and acceleration is a negative maximum. At 0.5 seconds displacement is zero (returned to the equilibrium position), velocity is a negative maximum and acceleration is zero. At 0.75 seconds displacement is maximum negative, velocity is zero and acceleration is maximum positive. The mass returns to the equilibrium position and has the original conditions of zero displacement, maximum positive velocity and zero acceleration.

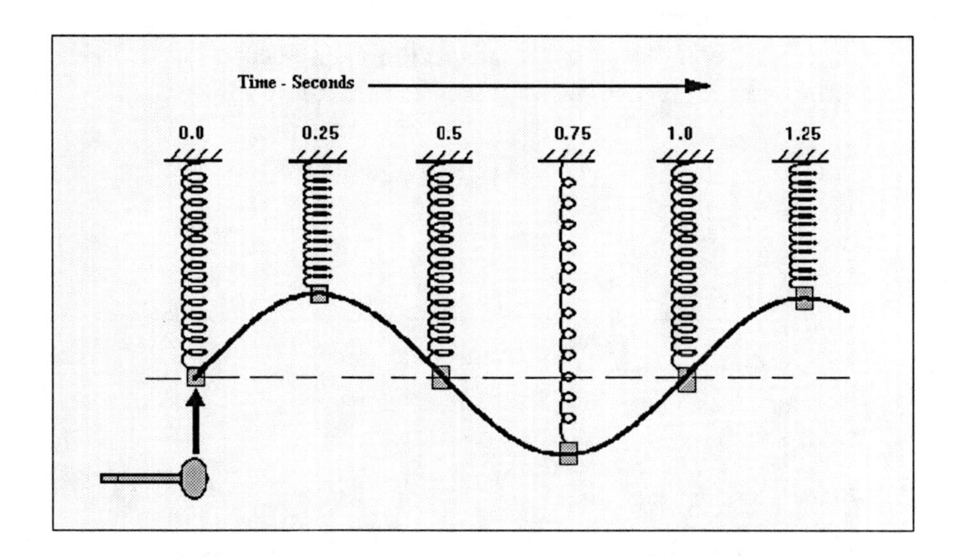


**Figure 1-2. The hammer impact is another way of exciting a mass and spring into free vibration. The pulse duration is so short compared to the time period of one vibration cycle that, for all practical purposes, the process begins with a condition of maximum velocity.**

Throughout this text the motion of masses or points on a vibrating structure will often be described by plotting either the displacement, velocity or acceleration versus time. Figure 1-3 is a plot of displacement versus time for the mass bouncing on the spring as a result of the hammer impact depicted in Figure 1-2. The Figure 1-3 plot follows the solid curve that traces the Figure 1-2 mass motion, extending the plot over several cycles of vibration. The vibration is plotted over a period of six seconds.

Two parameters are used in conjunction with the trigonometric sine function to describe the oscillatory motion: 1) Amplitude and 2) Frequency. The amplitude in Figure 1-3 refers to the maximum or peak displacement, which is 1.0 inch in this example. The frequency refers to the rate of oscillations, given in units called Hertz (Hz), or number of cycles per second. The oscillations in Figure 1-3 occur at a rate of one completed cycle of motion each second, or 1.0 Hz. The Figure1- 3 plot is also described by the algebraic formula:

$$Y(t) = Y_{max}\sin(2\pi\nu t) \tag{1-1}$$

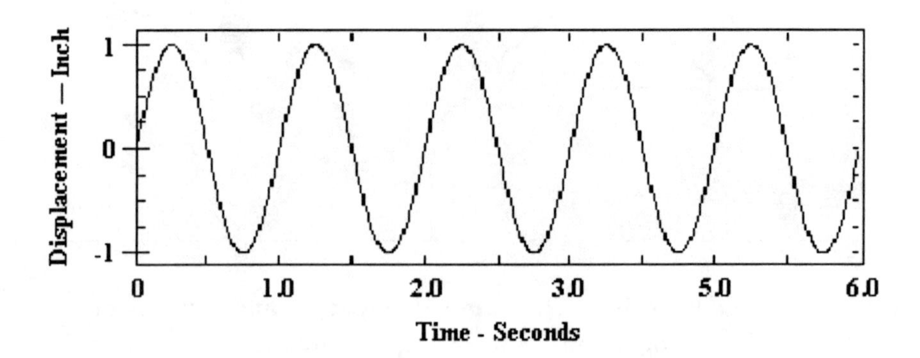


**Figure 1-3. Graph of displacement versus time for the vibration process illustrated in Figure 1-1.**

where Y is the instantaneous displacement at any time, t, $Y_{max}$ is the amplitude and ν (greek letter nu) is the frequency.

The motion represented by Figure 1-2 and Figure 1-3 is known as simple harmonic motion or sinusoidal motion. The sine function used in Equation (1-1) represents the displacement versus time trigonometric relation. Figure 1-4 helps clarify the trigonometric relation.

The left side of Figure 1-4 represents the motion of a point as it moves counter clockwise around the circumference of a circle whose radius is $Y_{max}$. The point progresses through a sequence of positions spaced 45 degrees apart. These points are numbered 1 through 8. Imagine that the point is continually moving with this circular motion and that it takes one second to move all the way around the circle once. Completing one trip around the circle will be referred to as one cycle of motion. The frequency or rate of moving around the circle is then one cycle per second.

Now, focus attention on just the vertical displacement of the point at each numbered position. The vertical line through the center of the circle in the figure is labeled as the Y axis, and the vertical displacement will be referred to as the value of Y. It is the projected position of the point onto the Y axis that actually represents the up and down vibration of the mass. The motion of the point around the circle is being used to assist the understanding of the way that the trigonometric sine function along with amplitude and phase angle represent the vertical oscillation.

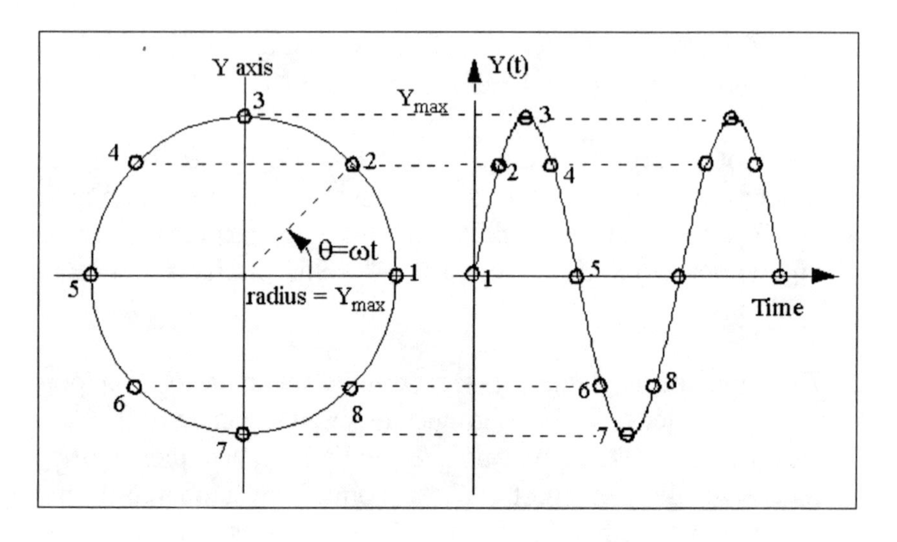


**Figure 1-4. A sequence of positions moving counter clockwise around a circle through angle, theta (θ), is matched to the corresponding positions on a graph of displacement versus time for a simple harmonic oscillator.**

Figure 1-5 constructs a right triangle with angle, θ (greek letter theta). Recall from trigonometry that the sine of the angle, θ, is defined as the ratio of the side opposite the angle (labeled as Y) divided by the hypotenuse (labeled as radius, Ymax).

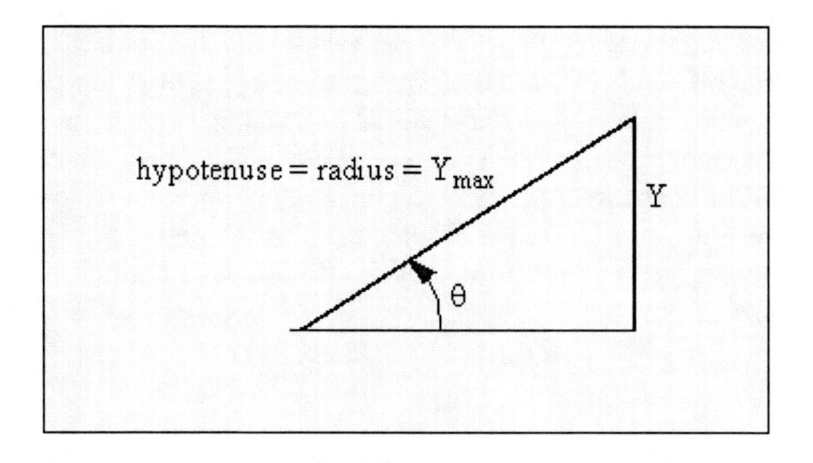


**Figure 1-5. A right triangle with defining parameters for the trigonometric sine definition.**

This is expressed by the equation:

$$\sin(\theta) = \frac{Y}{\text{radius}} \tag{1-2}$$

Or, using the value, $Y_{max}$, for the hypotenuse or radius,

$$\sin(\theta) = \frac{Y}{Y_{max}} \tag{1-3}$$

It is seen from equation (1-3) that the Y displacement value for any angle, $\theta$, and peak displacement or circle radius, $Y_{max}$, is:

$$Y = Y_{max}\sin(\theta) \tag{1-4}$$

Unless otherwise indicated, this text will always use the unit, radian, when referring to a variable angle, $\theta$. One radian of angle is that angle swept out when moving counter clockwise along the circumference of a circle such that the arc length along the circumference is equal to the radius of the circle. Since the circumference of a circle is known to be equal to $2\pi$ times the radius, then an angle of $2\pi$ radians would be swept out by moving around the complete circumference of the circle. This also implies that $2\pi$ radians correspond to 360 degrees. And since moving around the complete circumference of the circle, i.e., through 360 degrees or $2\pi$ radians, corresponds to one cycle of motion in Figure 1-4, it is seen that

$$1 \text{ cycle} = 2\pi \text{ radians} \qquad (1\text{-}5)$$

A special symbol, $\omega$ (greek letter omega), will be used to represent the rate of change of angle in radians with passage of time. Therefore, the angular position, $\theta$, of the point moving around the circle circumference can be represented at any time, t, as

$$\theta = \omega t \qquad (1\text{-}6)$$

Now, a formula for the vertical position of the point along the Y axis for any time, t, can be expressed:

$$Y(t) = Y_{max}\sin(\omega t) \qquad (1\text{-}7)$$

Since there are $2\pi$ radians per cycle (equation (1-5)), and frequency in Hz, $\nu$, is cycles per second, the relationship between frequency in radians per second and frequency in Hz is

$$\omega = 2\pi\nu \qquad (1\text{-}8)$$

Equation (1-7) can now be written using units of Hz (Hertz, cycles per sec), $\nu$:

$$Y(t) = Y_{max}\sin(2\pi\nu t) \qquad (1\text{-}9)$$

The application of the trigonometric sine function of equation (1-9) is the basis for referring to the curves of Figure 1-3 and right side of Figure 1-4 as sine waves. Matching the sequence of numbered positions between the circle on the left side to the sine wave on the right side of Figure 1-4, makes clear the use of the trigonometric sine function in the mathematical description of a sine wave.

The sine waveform is fundamental to the understanding of structural vibrations. It will be seen that even when the motion of a vibrating structure is complex, characterized by irregular movements, that motion can still be understood as a superposition of many different sine waves, each having different amplitudes and frequencies.

## 1.3 Velocity Sine and Cosine Waveforms

When describing the oscillating mass of Figure 1-2 we took note of the velocity at key points of the oscillation cycle. The velocity was a maximum at the start of the process, i.e., at the first instant the mass was set in motion by the hammer impact. Then the velocity slowed to zero at 0.25 seconds under the breaking reaction force of the stretching spring (the peak displacement position), changed to a negative maximum at 0.5 seconds (the equilibrium zero displacement position), decreased to zero velocity at 0.75 seconds (the maximum negative displacement position), and finally the completion of one cycle of oscillation at 1.0 second brought the mass back to the initial displaced position with the original maximum positive velocity. The time history of velocity is represented by the plot of velocity versus time in Figure 1-6.

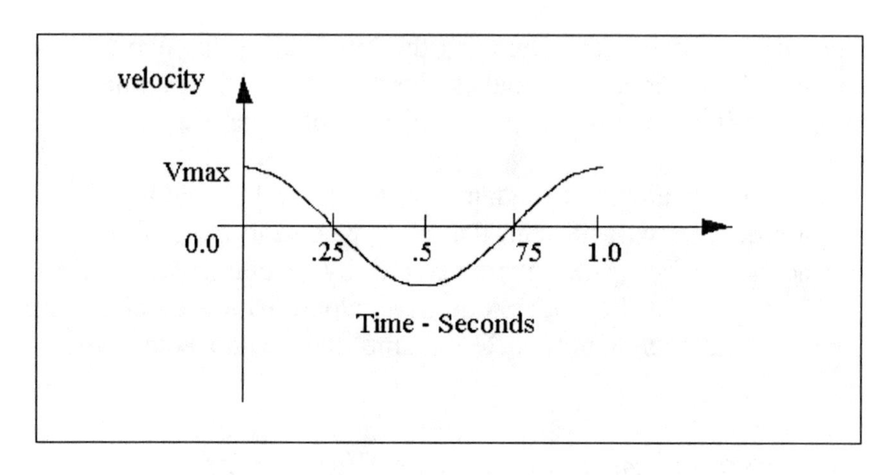


**Figure 1-6. Velocity versus time for Figure 1-2 vibration. The vibration when a hammer imparts an initial maximum velocity.**

A sine function may be used to represent the velocity versus time plot of Figure 1-6, but a phase angle must be added to the argument to account for the way velocity starts at t=0 with a maximum value. The phase of the velocity is shifted by 90°, or π/2 radians. The formula for the velocity, V(t), sine wave with phase shift is

$$V(t) = V_{max}\sin(\omega t + \pi/2) \quad (1\text{-}10)$$

The 90° phase shift for the circular motion of a point and the related phase shifted sine wave is shown in Figure 1-7. Note that the point motion begins at time, $t = 0$, at the angle, $\theta = \pi/2$ (90 degrees), rather than beginning at the angle, $\theta = 0$.

Of course this wave form is also known as a cosine function of time. The algebraic relationship between the cosine and sine functions is:

$$\cos(\theta) = \sin(\theta + \pi/2) \quad (1\text{-}11)$$

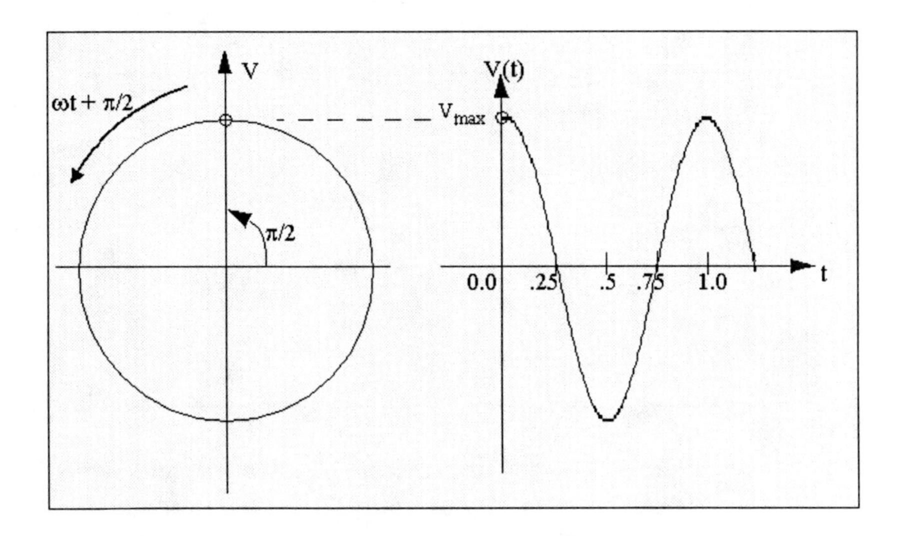


**Figure 1-7. Using a phase shift to allow a trigonometric sine function to describe the velocity wave form, which starts at a maximum value, $V_{max}$.**

Using the cosine definition avoids the need for a phase shift in the algebraic expression:

$$V(t) = V_{max}\cos(\omega t) \tag{1-12}$$

Or, using frequency, ν, in Hz, equation (1-12) is written

$$V(t) = V_{max}\cos(2\pi\nu t) \tag{1-13}$$

The displacement sine wave and velocity sine wave with phase shift (velocity cosine wave) are overlayed for comparison in Figure 1-8. The displacement sine function starts at a value of zero and the velocity phase shifted function starts at a positive maximum value. The wave forms are plotted over a time interval of three seconds and have a frequency of one Hz.

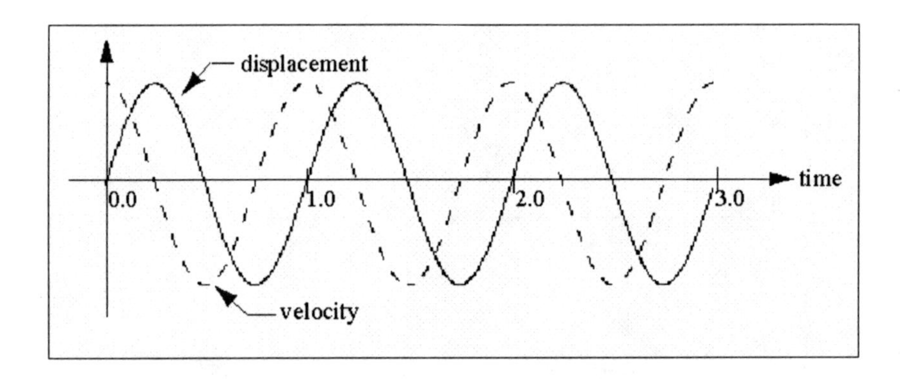


**Figure 1-8. Comparison of displacement and velocity wave forms for the simple harmonic motion of the mass bouncing on a spring in Figure 1-2.**

A triangle is sketched in Figure 1-9 to show the parameters used in trigonometry for defining a cosine function. The cosine is defined as the ratio of the side, V, adjacent to the angle, θ, divided by the hypotenuse, $V_{max}$ . Figure 1-10 compares the use of the trigonometric sine and cosine definitions for describing the velocity wave form as either a phase shifted sine wave or a cosine wave form with zero phase angle.

The definition of cosine for the Figure 1-9 sketch is:

$$\cos(\theta) = V/V_{max} \tag{1-14}$$

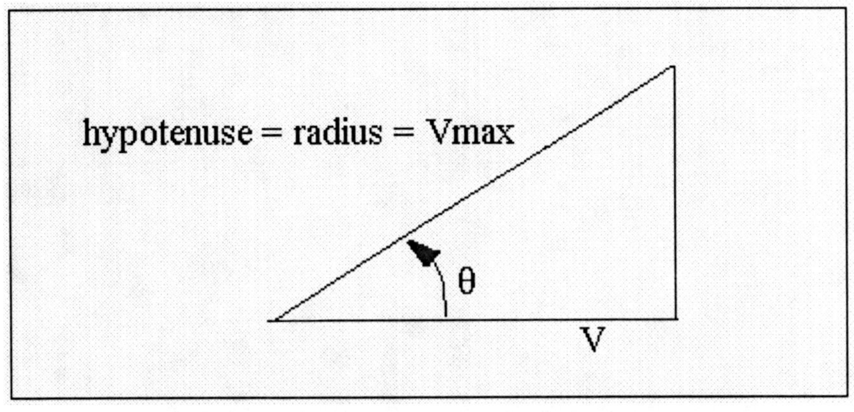


**Figure 1-9. Parameters used in defining cosine(θ): the ratio of the side adjacent to the angle (velocity, V) divided by the hypotenuse, $V_{max}$.**

The value of V may then be written as:

$$V = V_{max}\cos(\theta) \qquad (1\text{-}15)$$

The formula for the cosine description of the velocity wave form plotted in the lower left part of Figure 1-10 was written as equation (1-12) and is repeated here as equation (1-16):

$$V(t) = V_{max}\cos(\omega t) \qquad (1\text{-}16)$$

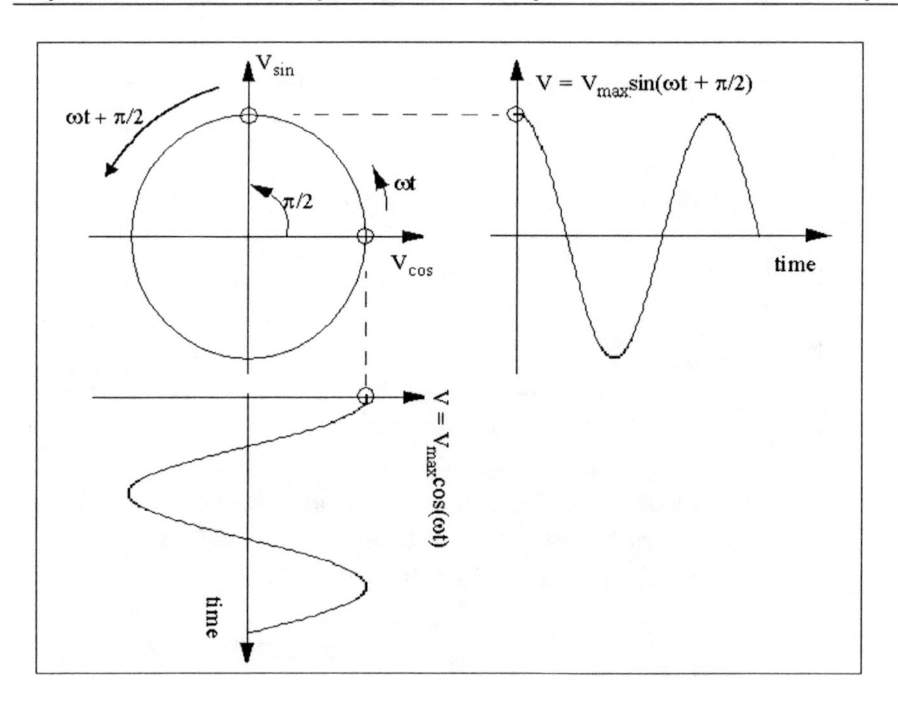


**Figure 1-10. A comparison of two ways of representing the velocity wave form. Upper right shows the phase shifted sine wave. The motion of the point on the circle begins at the π/2 phase position. The lower left plot is the cosine wave form. The point on the circle starts at the zero phase position.**

## 1.4 Acceleration Sine Waveform

The acceleration of the vibrating mass was noted at the key positions shown in Figure 1-2. At time equal to zero the acceleration was zero. When the displacement at 0.25 seconds was a positive maximum, the acceleration was maximum negative. At the 0.5 second equilibrium position the acceleration was zero. Acceleration was a positive maximum at 0.75 seconds when the position was a negative extreme. At the completion of one cycle the acceleration was again zero. The acceleration function of time, A(t), is a sine function, but the sign is the opposite of the displacement. The formula is

$$A(t) = -A_{max}\sin(\omega t) \qquad (1\text{-}17)$$

The acceleration versus time wave form for Figure 1-2 is plotted in Figure 1-11.

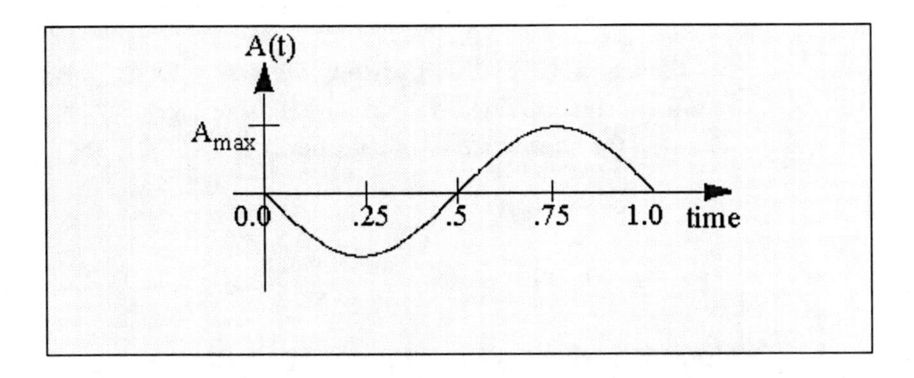


**Figure 1-11. Acceleration versus time for the oscillatory motion of the mass bouncing on the spring in Figure 1-2. The acceleration is described by a sine function with opposite sign from the displacement (180 degrees out of phase with displacement).**

The relationship between displacement, velocity and acceleration is summarized in Figure 1-12.

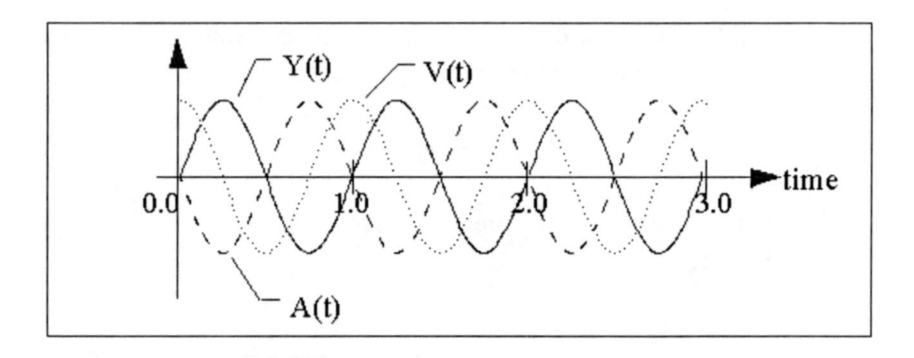


**Figure 1-12. Comparison of wave forms for displacement, Y(t), velocity, V(t), and acceleration, A(t), for simple harmonic motion.**

## 1.5 Velocity and Acceleration Amplitudes

We've been skirting the details of the actual velocity and acceleration amplitudes associated with sinusoidal displacement wave forms. The overlays in Figure 1-12 were not intended to properly reflect the actual numerical values of the amplitude relationships. The numerical values can be obtained from the velocity and acceleration formulas which will now be developed using calculus derivatives, beginning with a repetition of the displacement sine wave formula of equation (1-7),

$$Y(t) = Y_{max}\sin(\omega t) \qquad (1\text{-}18)$$

where $Y_{max}$ is the amplitude (maximum displacement) value.

The velocity, V(t), is the first derivative of displacement with respect to time, resulting in

$$V(t) = \omega Y_{max}\cos(\omega t) \qquad (1\text{-}19)$$

Or, using frequency, ν, in Hertz,

$$V(t) = 2\pi\nu Y_{max}\cos(2\pi\nu t) \tag{1-20}$$

Comparing this result, equation (1-20), with equation (1-13) the amplitude or maximum value of the velocity cosine wave form is related to the displacement amplitude and frequency as

$$V_{max} = 2\pi\nu Y_{max} \tag{1-21}$$

The acceleration function of time, A(t), results from taking the second derivative of displacement or the first derivative of velocity. Taking the first derivative of the velocity function of equation (1-19) gives

$$A(t) = -\omega^2 Y_{max}\sin(\omega t) \tag{1-22}$$

or, again using frequency, ν, in Hertz,

$$A(t) = -4\pi^2\nu^2 Y_{max}\sin(2\pi\nu t) \tag{1-23}$$

Comparing equation (1-23) to equation (1-17) shows the acceleration sine wave amplitude or maximum value, $A_{max}$, to be

$$A_{max} = 4\pi^2\nu^2 Y_{max} \tag{1-24}$$

Notice the squared frequency relationship between displacement and acceleration. This accounts for the often observed vibration lab phenomena in which a large displacement at a very low frequency is accompanied by a small acceleration level. At very high frequencies and small displacements, perhaps microinches, the acceleration can be quite high.

The units used in equation (1-20) and equation (1-23) for velocity and acceleration depend on the units being used for displacement. If displacement is expressed in units of inches, then velocity is in units of inches/sec and acceleration is in units of inches/sec/sec.

Acceleration is usually the parameter measured for characterizing the vibration of a structure. This is because of the difficulty of performing accurate measurements of displacement or velocity over a wide frequency range. Also, test article measurment locations are too often inaccessable to displacement and velocity measurement devices. Acceleration, on the other hand, is easily measured using devices called accelerometers. Test labs usually calibrate their accelerometer instrumentation channels to provide data in units of G's. An acceleration wave form, G(t), in units of G's is obtained by forming the ratio of acceleration, A(t), to the constant acceleration of gravity, $g_0$:

$$G(t) = \frac{1}{g_0} A(t) \tag{1-25}$$

where $g_0$ = 386.4 in/sec/sec. Writing equation (1-17) again, using G's and frequency, ν, in Hertz,

$$G(t) = -\frac{A_{max}}{386.4} \sin(2\pi\nu t) \tag{1-26}$$

Or,

$$G(t) = -G_{max} \sin(2\pi\nu t) \tag{1-27}$$

# Chapter II

# INPUT and RESPONSE VIBRATION

## 2.1 SDOF Response to Base Motion at Resonance

Free vibration of a simple spring-mass or SDOF (Single Degree Of Freedom) system was described in Chapter I. Two cases were considered: releasing the mass from an initial displacement and impacting with a hammer to produce an initial velocity condition. Now, a process will be considered in which sinusoidal motion is applied to the base to which the spring-mass system is attached. The motion of the mass responding to this base motion will be examined for the case that the applied base motion is a sine wave. Figure 2-1 diagrams the process.

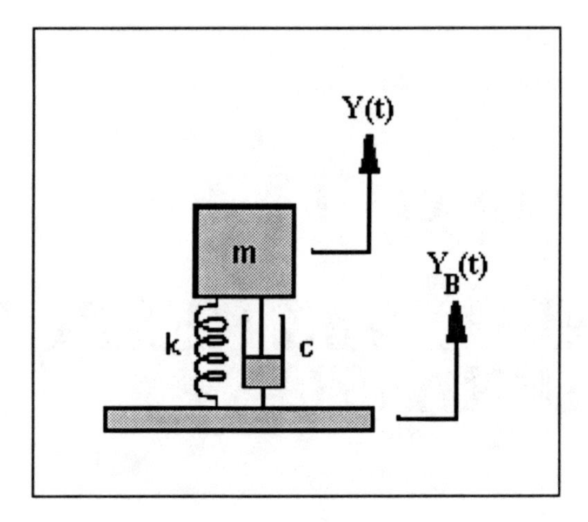


**Figure 2-1. SDOF (Single-Degree-Of-Freedom) system excited by input vibration at the base. The base motion is represented by the displacement versus time function, $Y_B(t)$, and the mass response motion is represented by the function, Y(t).**

A damper element has been included in the SDOF system to more accurately account for the behavior of real structures. The base is assumed to be an ideal infinitely rigid support structure whose motion can be controlled so as to produce any desired sinusoidal displacement of specified amplitude and frequency. We would like to understand the relationship between the response motion of the suspended mass and the input base motion.

The input motion at the base and response motion of the mass are overlaid as displacement versus time functions in Figure 2-2. At time equals zero a displacement sine wave of one inch amplitude is applied at the base. The applied frequency is one Hertz. This is also the resonance frequency of the SDOF system. Later, examples of exciting the system at frequencies other than resonance will be considered.

Recall from the two examples of Chapter I, Figures 1-1 and 1-2, that the system vibrated freely at a frequency of one Hertz when displaced and released or struck with a hammer. This condition of free vibration described in Chapter I is known as a resonant condition, and the frequency at which the system resonates is the resonance frequency. If the damper is ignored a very simple formula may be used to calculate the undamped resonance frequency, $\nu_r$, of a system with mass, m, and spring stiffness, k:

$$\nu_r = \frac{1}{2\pi}\sqrt{\frac{k}{m}} \qquad (2\text{-}1)$$

A more complicated formula that includes the damping could be used, but that frequency is usually only slightly different. Equation (2-1) can be derived from considerations of the free vibration process described previously in Figure 1-2. Under the conditions of free vibration of the mass on a spring (without damping), energy transfers from kinetic energy of the mass (when its velocity is a maximum, $V_{max}$ , and displacement is zero) to potential energy of the spring (when velocity is zero and the displacement is a maximum, $Y_{max}$). The equation for maximum kinetic energy, K.E., at maximum velocity, $V_{max}$, for mass, m, is

$$K.E. = \frac{1}{2} m {V_{max}}^2 \qquad (2\text{-}2)$$

And the potential energy, P.E., stored in the spring of stiffness, k, at maximum deflection, $Y_{max}$ is

$$P.E. = \frac{1}{2} k {Y_{max}}^2 \qquad (2\text{-}3)$$

Now, energy must be conserved when transferring between kinetic and potential, so

$$K.E. = P.E. \tag{2-4}$$

or

$$mV_{max}^2 = kY_{max}^2 \tag{2-5}$$

Rearranging equation (2-5),

$$\frac{V_{max}^2}{Y_{max}^2} = \frac{k}{m} \tag{2-6}$$

Substituting the right hand side of equation (1-21) for $V_{max}$ in the left side of equation (2-6),

$$\frac{4\pi^2 \nu^2 Y_{max}^2}{Y_{max}^2} = \frac{k}{m} \tag{2-7}$$

Solving for the frequency, where now the frequency is at resonance, i.e., $\nu_r$,

$$\nu_r = \frac{1}{2\pi}\sqrt{\frac{k}{m}} \tag{2-8}$$

Figure 2-1 represents a resonant system whose resonance frequency is one Hertz. When the mass-spring-dashpot is driven at its resonance frequency, the amplitude of the mass motion increases to many times the input motion at the base. The input and response motions are represented by the plots in Figure 2-2. The mass response motion builds up to a steady state level in about seven or eight seconds in this example.

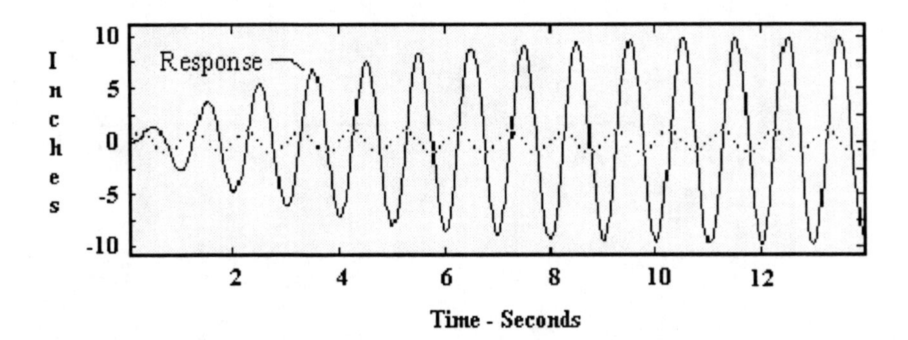


**Figure 2-2. Input base motion and response displacement for SDOF.**

After reaching the steady state level the mass response motion is sinusoidal with frequency exactly the same as the input base motion. A characteristic of any linear system is that, when excited into steady state response vibration, the system can only respond at whatever frequency or frequencies are applied as input vibration. However, there is a phase shift of approximately $\pi/2$ radians (90 degrees). Figure 2-3 zooms in on the Figure 2-2 data in the neighborhood of 10 seconds to provide a better view of the $\pi/2$ radian phase shift at steady state.

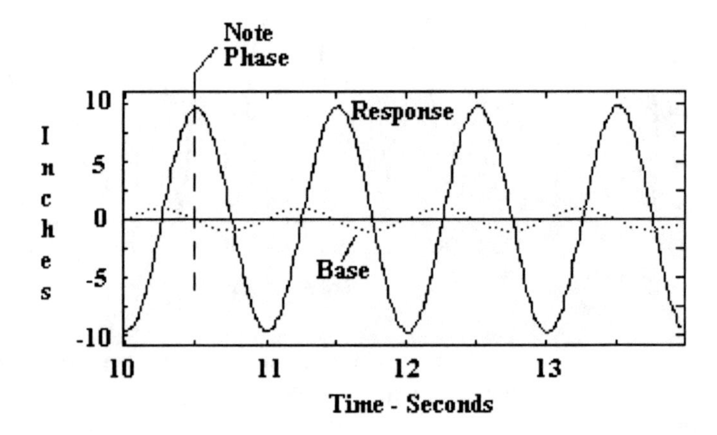


**Figure 2-3. Close-up of Figure 2-2 data showing a phase shift of approximately π/2 radians (90 degrees) for the response sine wave with respect to the input base motion. The response always lags the input by approximately π/2 radians when the system is lightly damped and is excited into steady state vibration at its resonance frequency.**

All lightly damped linear SDOF systems respond at resonance with a phase shift of approximately π/2 radians (90 degrees) with respect to the input sine wave. The system could be excited with vibration at frequencies other than the resonance frequency, and in that case there would still be a phase shift, but it would not be a π/2 radian phase shift and the response amplitude would not be as large.

The ratio of the response displacement to the base input displacement is called the amplification factor (sometimes it's called transmissibility). The size of the amplification factor ("amp factor") for a given SDOF system depends on the amount of damping associated with the damping element. It is not uncommon to encounter amplification factors of 40 to 50 for lightly damped structures when applying input sinusoidal motion at a resonance frequency of the structure. The amplification factor is seen in Figure

2-3 to be approximately 10. The input displacement is 1.0 inch and the response is approximately 10.0 inches.

## 2.2 Burst Sine Base Motion

Another type of base motion is burst sine. A displacement sine wave is applied to the base as before, except now the excitation is abruptly terminated after some relatively short time, allowing the system to continue vibrating freely. But with damping present the vibration will decay. This is in contrast to Figures 1-1 and 1-2 where the free vibration continued indefinitely without any damping in the system. The rate at which the free vibration decays depends on the amount of damping in the system. Figure 2-4 overlays plots of the input base motion burst sine function and the response motion. The input sine wave is applied with a frequency of 1 Hz (the system resonance frequency) for 10 seconds and then removed. The response motion has increased to a steady state amplitude of 10 inches by then, but within the next eight seconds, following removal of the input base motion (the base is suddenly clamped so it cannot move), the response amplitude has reduced to about 1 inch as a result of the damping.

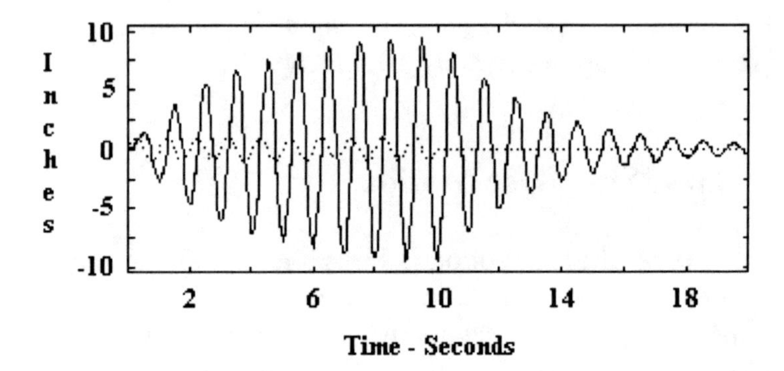


**Figure 2-4. SDOF response to base motion burst sine. The input sine wave is terminated at 10 seconds. The system continues to vibrate freely but decays due to the presence of damping.**

A closer look at the decay portion of the Figure 2-4 plots is shown in Figure 2-5.

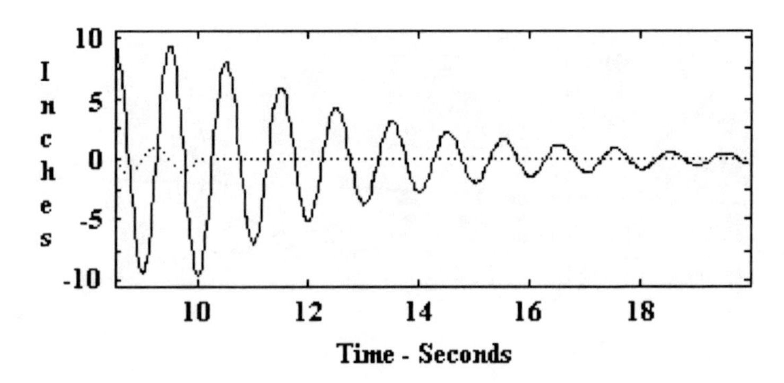


**Figure 2-5. Expanded view of the decay portion of the response vibration. The base motion burst sine function has been stopped at 10 seconds.**

Figure 2-6 compares the response of a heavily damped system to the more lightly damped results of Figure 2-4. The same burst sine function is applied to the base, but now the amount of damping has been increased. The response builds up to the steady state level quicker than before and the response amplitude is much less, about 0.4 inch. When the input vibration is removed at 10 seconds the response decay is much quicker.

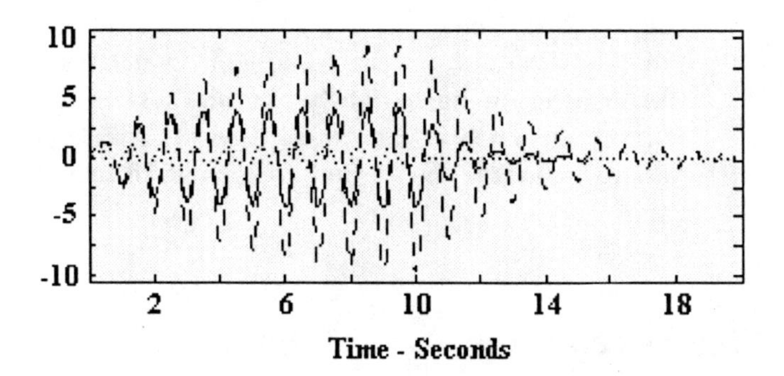


**Figure 2-6. Heavily damped response to burst sine (solid curve) compared to the previous more lightly damped response of Figure 2-4.**

Figure 2-7 zooms in on the decay portion of the Figure 2-6 plots.

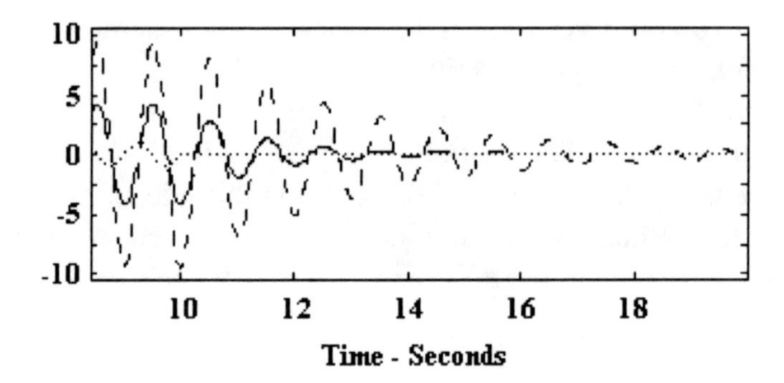


**Figure 2-7. Decay portion of Figure 2-6 plots. The heavily damped system (solid curve) decays much faster than the more lightly damped system (dashed curve). The input amplitude (dotted curve) abruptly goes from 1.0 inch to zero at 10 seconds.**

## 2.3 Sine Sweep Input Base Motion

A process familiar to every vibration test lab engineer or technician is the input sine sweep vibration test. Most vibration test labs are equiped to perform the test routinely. An electrodynamic or hydraulic shaker is controlled by a computer to produce a sine wave that begins at a specified frequency, say 5 Hz, and continuously increases the frequency, perhaps sweeping up to 2000 Hz before terminating the test.

The details of controlling the base motion (shaker table motion) become quite involved because of the stringent specifications usually imposed on the control of the input sine wave. The amplitude must be controlled to a desired level within a specified tolerence, and the sweep rate (rate of increasing frequency with time) must also be controlled within a specified tolerence. In practice the input motion is usually specified in G's (acceleration). Controlling the acceleration to a constant amplitude while sweeping through a wide frequency range is a process that is easy to conceive but difficult to implement. Test labs procure commercially available shaker systems with instrumentation, computer with appropriate signal generation and data acquisition electronics and software designed to perform the sine sweep vibration test.

Again we refer to Figure 2-1, this time illustrating the application of a swept sine wave as the input base motion. The amplitude is held constant at one inch peak displacement throughout the sweep. A plot of the first ten seconds of the sweep, increasing frequency from zero to 10 Hz, is shown in Figure 2-8 below. Note the sine wave peaks crowding closer together as time progresses and the frequency gets higher.

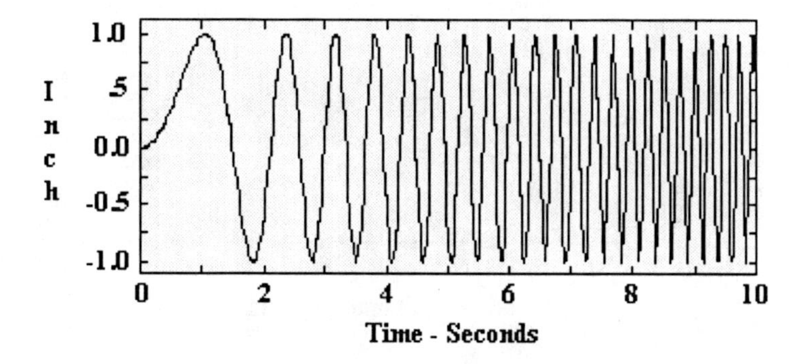


**Figure 2-8. Plot of input base motion sine sweep. Just the first 10 seconds of data are shown. Frequency continually increases over this period.**

Figure 2-9 overlays plots of the SDOF response (solid curve) compared to the input constant amplitude base motion. Notice that the response amplitude builds up to a large amplitude as the input frequency passes through the system resonance. In fact the system is so lightly damped that the resonance frequency is still damping out long after the input frequency has swept past resonance. The response of the SDOF system to frequencies toward the high end of the sweep are only a fraction of the input base motion (portion of curve labeled "minimal response").

Figure 2-10 provides a more clear picture of the input and response during the first ten seconds of the sweep. Actually, the sweep rate for this example is too fast for the SDOF system to reach a momentary steady state level at resonance. The ratio of response motion to input motion for the highest response amplitude is only 4.0 at 3.4 seconds. Compare this to the amp factor of 10.0 for the steady state response of the SDOF as seen in Figure 2-2.

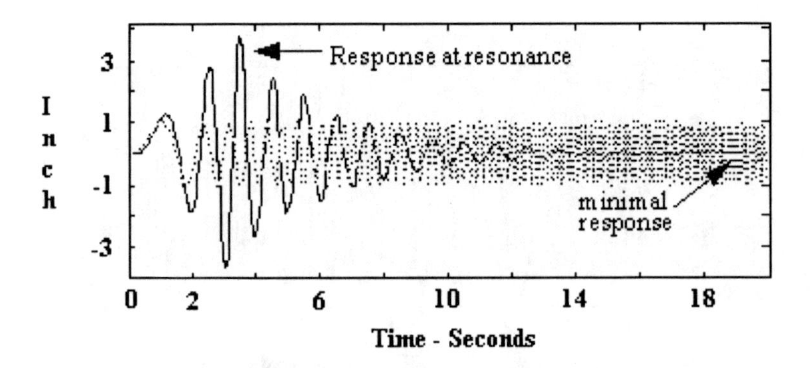


**Figure 2-9. Comparison of response displacement versus time (solid curve) to the input base motion sine sweep plot (dotted curve). The SDOF response peaks up as the input passes through its resonance frequency.**

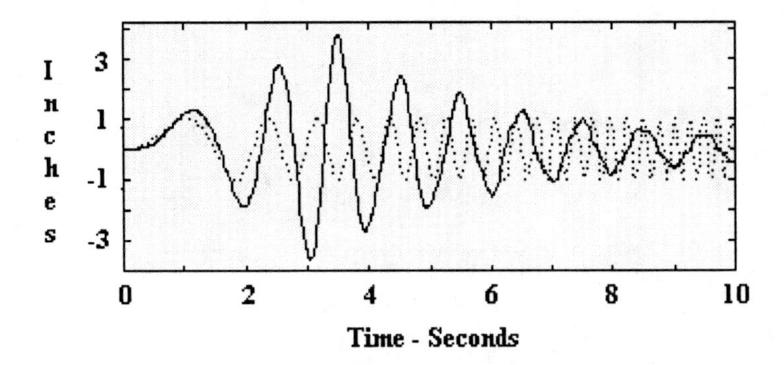


**Figure 2-10. A closer look at the SDOF response to a sine sweep input for the first ten seconds.**

To illustrate the effect of allowing more time for the system to respond in the resonance region, another example is plotted in Figure 2-11 using a slower sweep rate. The input amplitude is still maintained at 1.0 inch, and now the response motion is the same as the input base motion at frequencies much lower than resonance, but the ratio of response to input in the resonance region gets up to 7.0. The resonance response is still not as high as steady state response, but it is significantly higher than that of the faster sweep rate.

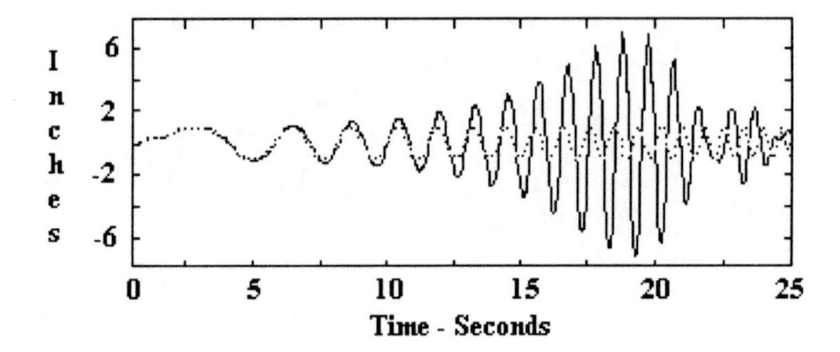


**Figure 2-11. Response to a slower sweep rate. The SDOF motion is the same as the input base motion for low frequencies, before encountering the resonance frequency region. Also, the slower sweep allows time in the region of resonance for the system to momentarily build up to a higher amplitude than was seen with the faster sweep rate of Figure 2-10.**

Now, look at the SDOF response at frequencies above the resonance frequency. Figure 2-9 demonstrates that there is a very minimal response to base motion when the input frequency is much higher than the resonance frequency of the system. Figure 2-12 zooms into the end of the Figure 2-9 time period to provide a detailed view of the response plot.

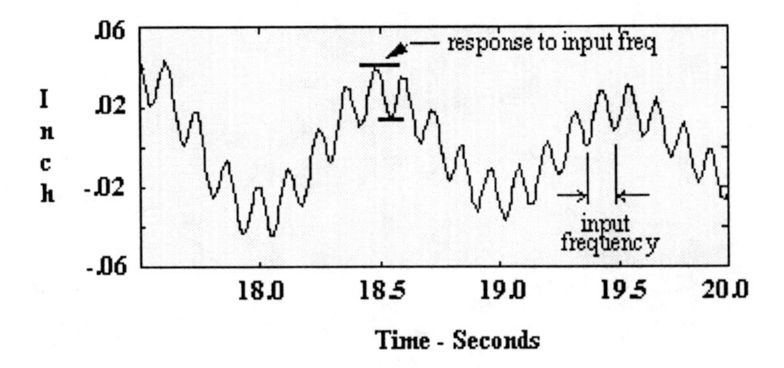


**Figure 2-12. Plot of response motion zoomed into the 17.5 to 20.0 second range of Figure 2-9. Even though the input base motion has an amplitude of 1.0 inch, the system is responding with about 0.015 inch amplitude at this relatively high frequency (high compared to the 1 Hz resonance frequency). The system resonance is still in the process of decaying away, and at this point its amplitude is even larger than the direct response to the current input frequency.**

This shows just how small the response amplitude is at frequencies well above resonance.

Again, there are two frequencies evident at this point in the sweep. The system is responding directly to the frequency of the input base motion at this instant, but the amplitude associated with the resonance frequency is still in the process of damping out. The amplitude of the resonance is actually greater than the amplitude due to the base motion frequency at this point.

At very high frequencies the SDOF system is considered to be isolated from the base vibration. In fact this is the principle of vibration isolation used in design practice to protect equipment from input vibration. Figure 2-13 illustrates the principle. Vehicle bodies are mounted to the wheel and axle assembly through suspension springs and dampers (shock absorbers).

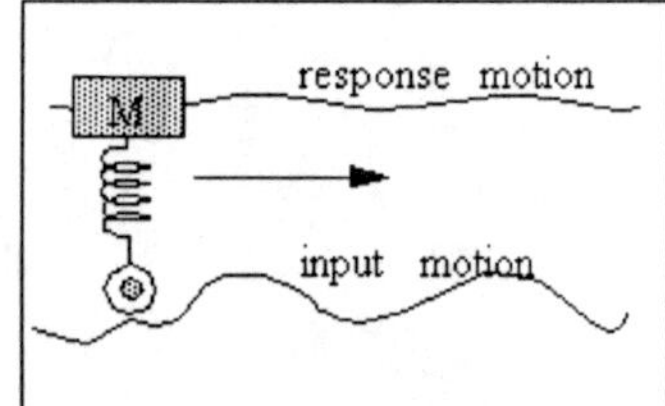


**Figure 2-13. Vibration isolation from the input road motion is achieved with a suspension system design that incorporates a spring and damper. Input road displacement is isolated at the vehicle body mass if the input frequencies are high compared to the resonance frequency of the spring-mass system.**

As long as the frequencies of vibration associated with the road profile are much higher than the resonance frequency of the vehicle mass-spring system, the vehicle body will be isolated from the road motion.

## 2.4 Amplification Factor Versus Frequency

The need to represent the response vibration relative to input base motion is quite common. Rather than plotting the time domain swept sine data engineers prefer to simply plot the amplification factor (ratio of sine wave response amplitude divided by the input sine amplitude) versus frequency. This plot is also referred to as a transmissibility plot and is usually derived from measurements of acceleration. The ratio will be the same whether derived from displacement, velocity or acceleration. This is seen by noting that frequency cancels in numerator and denominator

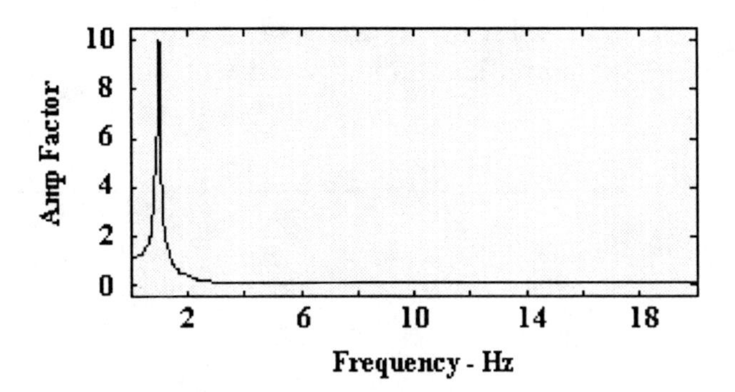


**Figure 2-14. Response characteristics for the SDOF system of Figure 2-1. Sine sweep data are plotted as the ratio of response amplitude to input amplitude, often referred to as a transmissibility or amplification factor plot.**

of response-to-input ratios for velocity and acceleration in the equations below. Using the velocity formula of equation 1-21, the amp factor for response velocity to input velocity is

$$V_{resp}/V_{input} = 2\pi\nu Y_{resp}/2\pi\nu Y_{input} \tag{2-9}$$

or, cancelling out the $2\pi\nu$ in numerator and denominator,

$$V_{resp}/V_{input} = Y_{resp}/Y_{input} \tag{2-10}$$

Using equation 1-24 for acceleration, a similar result is obtained.

$$A_{resp}/A_{input} = 4\pi^2\nu^2 Y_{resp}/4\pi^2\nu^2 Y_{input} \tag{2-11}$$

or

$$A_{resp}/A_{input} = Y_{resp}/Y_{input} \qquad (2\text{-}12)$$

Figure 2-15 is a replot of Figure 2-14, representing a more popular form of plotting amplification factor data. It plots the $\log(Y_{resp}/Y_{input})$ versus $\log(\nu)$.

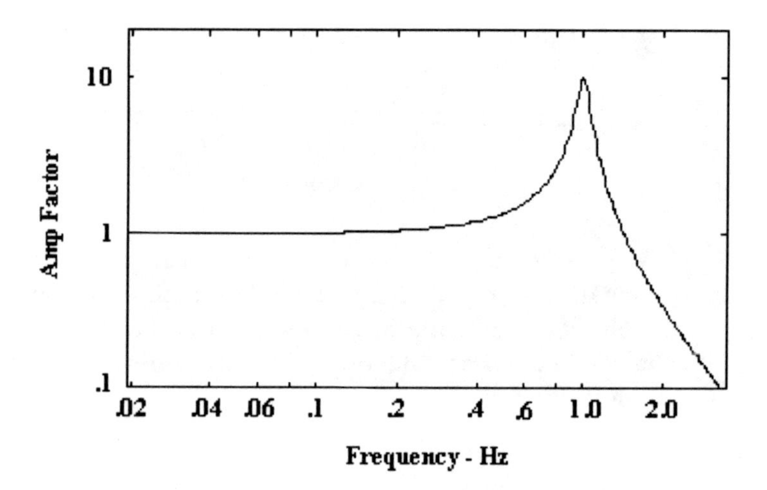


**Figure 2-15. The amplification factor (transmissibility) plot for the SDOF system of Figure 2-1. The log-log form is often preferred.**

## 2.5 SDOF Response Phase Angle

We have commented before on the -90 degree ($-\pi/2$ radians) phase shift of the response sine wave relative to the input sine wave when a SDOF is being excited at its resonance frequency. When performing the sine sweep test on the SDOF system of Figure 2-1, the phase angle begins at zero and gradually decreases as the frequency approaches the region of resonance, then moves through -90 degrees rather rapidly as frequency passes through resonance. The phase approaches -180 degrees ($-\pi$ radians) as the frequency sweeps to higher and higher frequencies. In fact the phase angle never quite reaches -180 degrees, no matter how high the frequency gets. The phase approaches -180 degrees as the frequency approaches infinity.

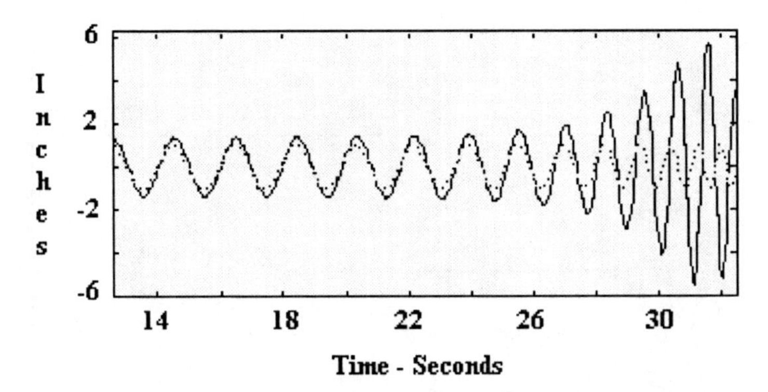


**Figure 2-16. Response to sine sweep using a logarithmic sweep rate. More time is spent at the lowest frequencies allowing a better look at the gradual phase shift before reaching the resonance frequency.**

The gradual shifting of the response phase angle as the input frequency approaches the system resonance may be recognized in Figure 2-16 where this time the sine sweep is performed using a logarithmic sweep rate. Sweeping with a logarithmic rate results in more time at the lower frequencies so that more low frequency cycles are available for inspection. The sine wave peaks in the response curve gradually shift to the right with each cycle, shifting to the -90 degree position at the resonance frequency. The shifting of response peaks to the right in the plot corresponds to more and more delay in the start of each new response cycle. This is a lag in time or phase lag, and the phase angle takes on an increasingly negative value.

As with the amp factor, the phase characteristics are best represented by a plot of phase versus frequency. The phase angle is plotted as a function of frequency along with amp factor in Figure 2- 17 below.

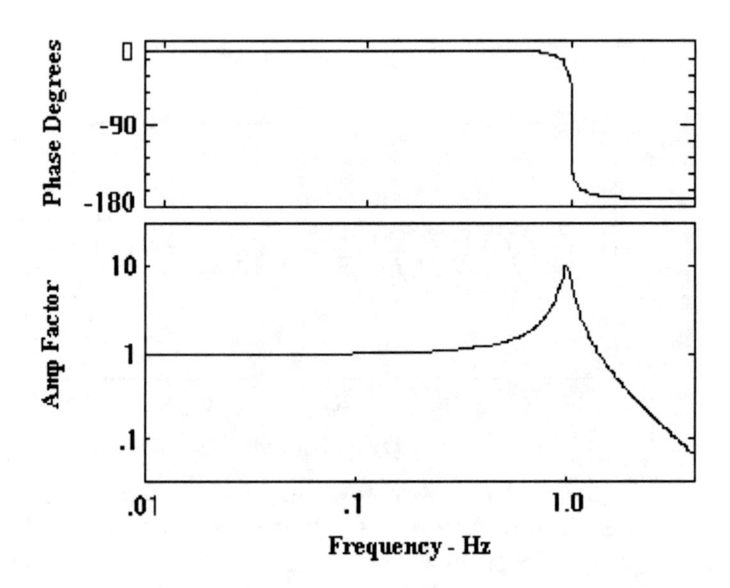


**Figure 2-17. The response phase angle (upper plot) for the SDOF system of Figure 2-1 is shown together with a plot of the log amp factor versus log frequency. The phase angle passes through -90 degrees as the sweeping frequency passes through resonance.**

# CHAPTER III
# SIGNAL PROCESSING

The vibration behavior of a simple SDOF has been described in the first two chapters using various examples. The behavior has been represented with equations and with time domain and frequency domain plots of vibration response. Figure 2-15 presented the Amp Factor as the response to input sine sweep base motion. Now, we look at what is involved in the measurement and processing of data so that graphs like Figure 2-15 may be obtained experimentally.

A transducer is a device that is capable of sensing some physical phenomena such as motion. It generates an electrical signal (a voltage varying with time) that has a known mathematical relation to a physical parameter such as displacement, velocity, or acceleration. The accelerometer is a transducer that enjoys wide use in the measurement of vibratory motion. It produces a time varying voltage that is directly proportional to the measured time-varying acceleration.

Signal processing is a subject concerned with the details of getting voltage signals into a computer and processing them into some desired form or result. You may be interested in looking at columns of numbers. A table might be used to compare numbers obtained from various transducer signals. You may be interested in the statistics computed from a column of numbers. Or, you may wish to display data graphically.

Data of interest in vibration applications often represent either time domain or frequency domain phenomena. Of particular interest later on in this text will be the processing of data to obtain the resonance frequencies and mode shapes of a vibrating structure. Much of the signal processing theory discussed here will be focused on aspects of data presentation in the frequency domain.

## 3.1 Sampling Theory

Getting a transducer signal into a computer means sampling one data point at a time so that the resulting sequence of numbers represents the time varying signal to good approximation. With a little reflection one realizes that even though a vibrating component involves continuously varying oscillatory displacement, there can never be a continuous data set of displacement versus time stored in a computer.

There can only be a sequence of *discrete* displacement values, *sampled at discrete points* in time. Ultimately, discrete displacement values are stored individually in computer memory cells. There is always some time separation, however small, between successive measured displacement values. We refer to these as discrete time domain data points. Figure 3-1 depicts this situation. The upper graph depicts an actual continuously varying displacement versus time record. The lower graph presents the set of discrete data points that have been sampled from the continuously varying displacement.

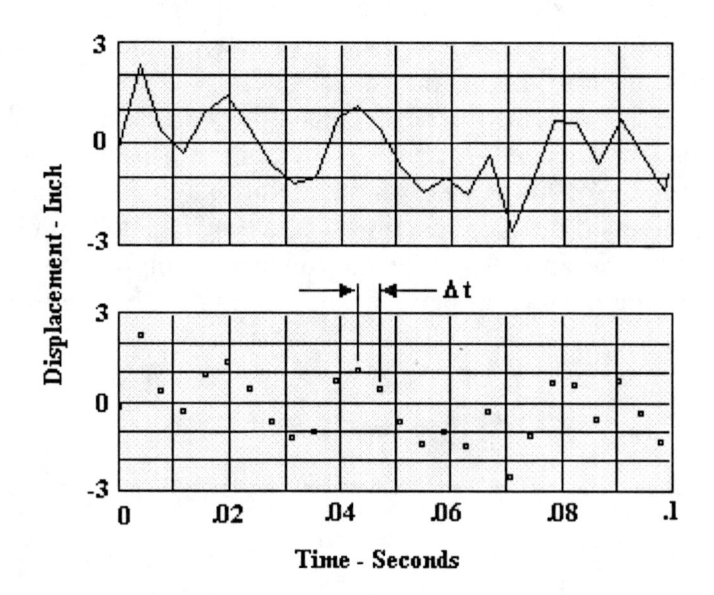


**Figure 3-1 Discrete sampling of data for computer processing. The upper displacement vs. time graph represents the measured continuous analog signal generated by a transducer. The lower plot indicates the discrete sampling. The sampling time increment, Δt, is about four milliseconds.**

The sampled data points are separated in time by some time increment, Δt (delta time). This defines the rate that data is being sampled into the computer:

$$\text{Sample Rate, SR} = (1 \text{ sample})/\Delta t$$

That is,

$$SR = 1/\Delta t \tag{3-1}$$

If the displacement is varying quite rapidly, then the Δt value must be very small to provide adequate time resolution. Another way of saying this is that the Sample Rate, SR, must be very fast in order to keep up with the rapidly changing displacement. In fact, how fast the displacement is fluctuating places a requirement on the Sample Rate for faithful reproduction of the displacement signal. We can see quite clearly the way this works by considering an example displacement signal that is fluctuating between positive and negative values at 1000 cycles per second. Figure 3-2 graphs a sine function having a frequency of 1000 Hz and amplitude of one inch. In this example we have sampled the sine function at a rate of 10,000 Samples/Sec (10K S/Sec). This appears to be an adequate sampling for purposes of identifying the data as being a sine function with a frequency of 1000 Hz.

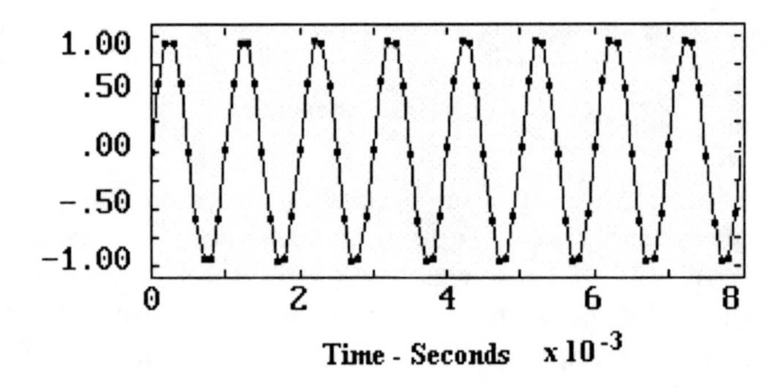


**Figure 3-2 A 1000 Hz sine function time series sampled at 10,000 samples/sec. The time samples are spaced at .0001 sec increments.**

There arises the question, "How low a sample rate could be used and still be able to identify the data as a sine function with a frequency of 1000 Hz?" The answer is 2000 S/Sec. In sample theory it is understood that the maximum measurable sine function frequency available from sampled data is a frequency that is one-half of the Sampling Rate. If the maximum frequency of interest is 1000 Hz, then you must sample at a rate that is at least twice this rate, i.e., 2000

Hz. This can be understood with the help of the graphs of Figure 3-3 and Figure 3-4.

Figure 3-3 illustrates the sampling of a 1 KHz (1000 Hz) sine signal using the minimum possible sampling rate that would be required for this signal. The sampling rate in this case (2000 samples/sec) is exactly twice the frequency of the actual sine function, and the time points are spaced at .0005 second increments. Each sampled data point falls on either a peak or a valley (+1 inch or –1 inch) of the displacement sine function. These are the discrete data points that would be stored in computer memory. Checking equation (3-1) using the given sample rate and time increment,

$$2000 \text{ S/Sec} = (1 \text{ Sample})/(.0005 \text{ Sec}) \qquad (3\text{-}2)$$

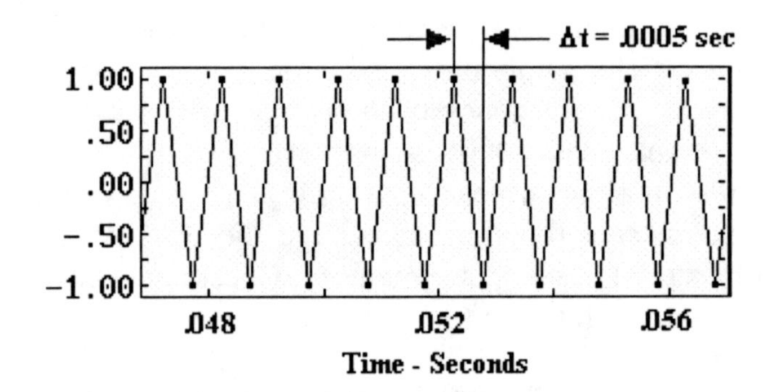


**Figure 3-3 A sine function having a frequency of 1 KHz (1000 Hz), sampled at exactly twice the sine frequency. The data are sampled at 2000 samples/sec at time points spaced at .0005 sec increments. This is the lowest sample rate possible for the replication of a 1 KHz sine function.**

Now the question arises as to what happens if the displacement signal is undersampled. What if the sample rate is less than twice the actual sine signal frequency? The result of undersampling (sampling too slowly) is shown in Figure 3-4.

Figure 3-4 presents an example in which the transducer signal having a frequency of 1 KHz (dotted curve) is sampled at just 775 S/Sec (solid curve with box markers at the sampled points). The lower plot zooms in over a shorter time duration of just a little more than eight milliseconds.

A close look at the undersampled plot reveals a slightly bazaar result. The resulting plot is actually that of another sine function having the same displacement amplitude (one inch) but a much lower frequency, approximately 225 Hz. This lower frequency resulting from the undersampling is known as an alias frequency. The notion here is that rather than produce the true frequency (1000 Hz in this case), the undersampled plot displays a substitute name, or alias name, for that frequency (225 Hz in this case). In fact it is generally the case that for any undersampled sine signal, there will result a different sine plot having an alias frequency, $\nu_A$, that is equal to the difference between twice the, undersampled sample rate, $SR_U$, and the true sine frequency, $\nu_s$:

$$\nu_A = (2 \text{ x } SR_U) - \nu_s \qquad (3\text{-}3)$$

It is obvious that undersampling a sine function has the potential for gross corruption of rapidly fluctuating vibration signals, undermining any attempt to properly interpret the vibration process under study.

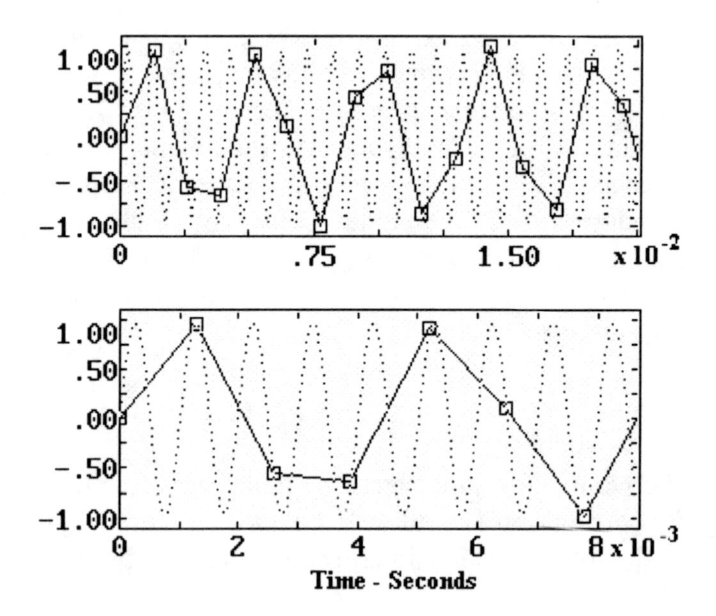


**Figure 3-4 Example result of undersampling a 1 KHz sine signal data set. The dotted curve is the 1 KHz signal, and the solid curve (with descrete data point markers) shows the result of using a sample rate that is too low. The lower plot zooms in on the first eight milliseconds of data. The undersampled data is said to be "aliased."**

Typically, a vibration signal will be comprised of a superposition of many different sine functions having different amplitudes and phase angles. The signal may be acquired with a sample rate that properly samples most of the frequencies but undersamples the higher frequencies in the signal. If the higher frequencies are of no interest, then an analog low pass filter may be used to remove the high frequencies from the signal before it is sampled. That is, all frequencies above one-half the sampling rate would be filtered out. Such a filter is called an anti-aliasing filter. Having removed all higher frequencies ahead of time, the final processed data will contain no erroneous alias frequencies.

## 3.2 Superposition of Sine Functions

Consider the waveform that results when two sampled sine functions of different amplitudes and frequencies are summed together. When we speak of summing two sampled sine functions, we are thinking of adding the pairs of data points at each time increment, or point in time, $t_n$, where n = 0,1,2,3, ..., N-1, and N is the total number of data points sampled into the computer.

A composite function, Y(t), results from the summation of two sine functions, $Y_1(t)$, and $Y_2(t)$, each function having a different frequencies $\omega_1$ and $\omega_2$:

$$Y(t) = Y_{1max}\sin(\omega_1 t) + Y_{2max}\sin(\omega_2 t) \qquad (3\text{-}4)$$

The data would be sampled into a computer at the discrete time points, $t_n$. For each point in time the composite data set has the displacement value,

$$Y(t_n) = Y_{1max}\sin(\omega_1 t_n) + Y_{2max}\sin(\omega_2 t_n) \qquad (3\text{-}5)$$

The time value corresponding to the nth data point in the set would be (using the sampling time increment, $\Delta t$):

$$t_n = n \times \Delta t \qquad (n = 0, 1, 2, 3, \ldots, N) \qquad (3\text{-}6)$$

Or, using the sample rate definition, $SR = 1/\Delta t$, of equation (3-1):

$$t_n = \frac{n}{SR} \qquad (3\text{-}7)$$

Figure 3-5 illustrates the summation of two sampled sine function data sets.

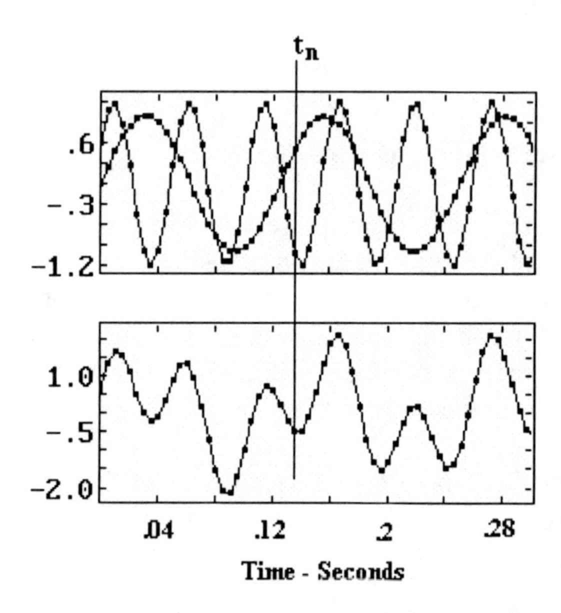


**Figure 3-5 The summation of two sine function data sets. The lower waveform results from adding a pair of data points at each discrete time point. The time-synchronized points at some time, $t_n$ (for some nth point), are marked.**

## 3.3 Fourier Analysis

One of the most powerful and popular operations in vibration analysis is the decomposition of the most arbitrarily complicated waveform into a set of sine functions of different frequencies and phase angles. This is the exact reverse of the superposition of sine functions illustrated in Figure 3-5. You start out with a complicated waveform such as the lower plot of Figure 3-5 and decompose it into the two separate sine functions overlaid in the upper plot of Figure 3-5.

One way of presenting the result of this type of analysis, or decomposition, is to plot the amplitudes of each sine function as a function of frequency. Figure 3-6 below plots the amplitudes of the two sine functions at their respective frequencies.

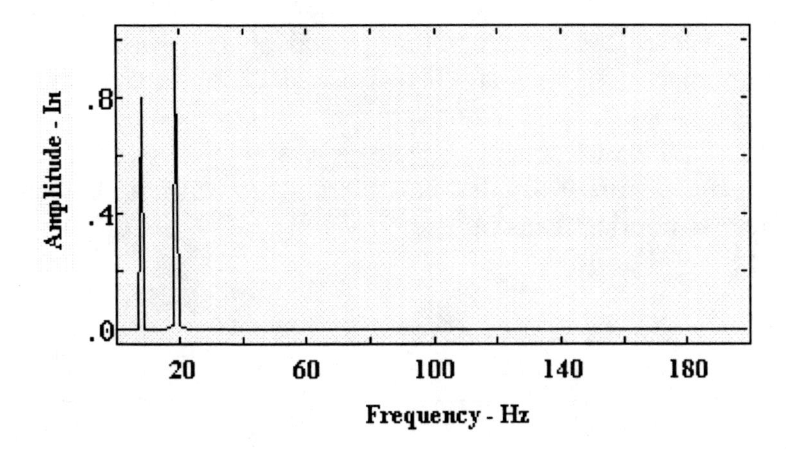


**Figure 3-6 Result of decomposing the complex waveform of lower Figure 3-5 into separate sine functions. The amplitudes of each of the two sine functions (0.8 In and 1.0 In) are displayed at their discrete frequency values of 8 Hz and 19 Hz. This result is known as a Fourier analysis, or FFT (Fast Fourier Transform) of the original time series waveform.**

Considerable effort has been expended over the years to develop efficient computer algorithms for the decomposition of any arbitrary waveform into its frequency spectrum. This is called Fourier analysis, and a landmark algorithm known as the FFT (Fast Fourier Transform) was developed by Cooley and Tukey about 1964. The algorithm executes very quickly (the order of one or two milliseconds for a set of 1024 time domain points) and takes advantage of certain trigonometric relationships for time domain data sets having a number of data points equal to some integral power of 2 (32 points, 64, 128, 256, 512, 1024, etc.). A straightforward discrete Fourier Transform executes more slowly, but performs the analysis for time series data sets having any number of data points (the number of data points in the time series, N, is not restricted to powers of 2).

The Inverse Fourier Transform accepts a set of frequency domain data, such as Figure 3-6, and sums up sine functions in the time domain represented by the frequency domain amplitudes and frequencies. As an example, we add another sine function having amplitude of 1.2 Inch at 35 Hz to our frequency spectrum of Figure 3-6. Then, we Inverse Transform back to the time domain. Figure 3-7 shows the new frequency spectrum after adding a new sinewave at 35 Hz, and Figure 3-8 is the new complex waveform obtained after the Inverse Fourier Transform.

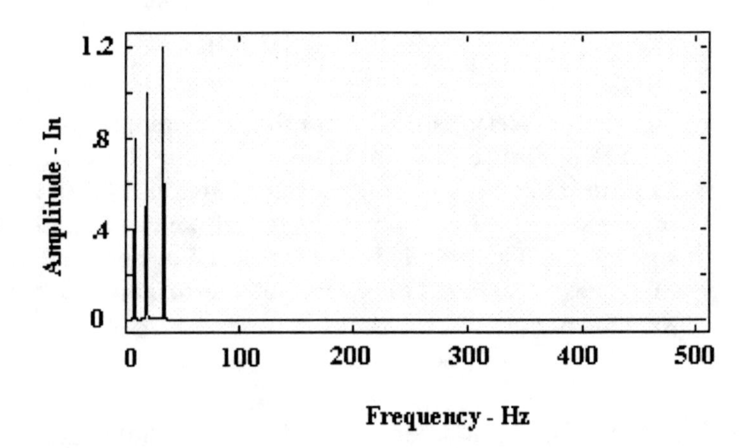


**Figure 3-7 A third sine function has been added to the frequency spectrum of Figure 3-6. The new sine amplitude is 1.2 Inch at a frequency of 35 Hz. This additional sine function was added by simply editing the frequency domain data of Figure 3-6.**

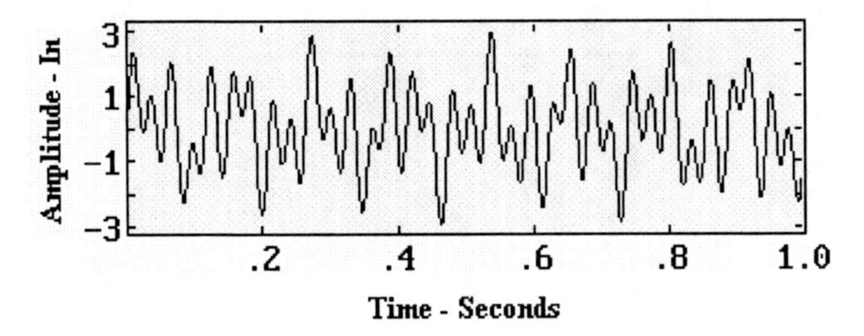


**Figure 3-8 The frequency spectrum of Figure 3-7 has been Inverse Fourier Transformed back into the time domain. The new waveform is now made up of the summation of three sine functions having frequencies of 8 Hz, 19 Hz, and 35 Hz.**

The frequency domain function (Fourier spectrum) of Figure 3-7 and the time series function of Figure 3-8 are known as Fourier Transform pairs. If you had been given the Figure 3-8 time series waveform initially, you could obtain the Figure 3-7 frequency domain function by performing a FFT. Conversely, the Figure 3-8 time domain function is obtained by performing an Inverse Fourier Transform on the frequency domain function of Figure 3-7.

Some signal processing software allows the FFT operation to be performed on a stored time domain data set by simply clicking on an FFT menu option. The computer program user need not know any of the underlying mathematics to implement such a sophisticated operation.

### 3.4 Accommodating Phase Angle with the FFT

A detail not to be overlooked is the way that phase angle is handled in Fourier Analysis. To understand this, we first observe that a cosine wave and a sine wave of the same frequency may be summed to yield a cosine wave having a phase angle as illustrated in Figure 3-9.

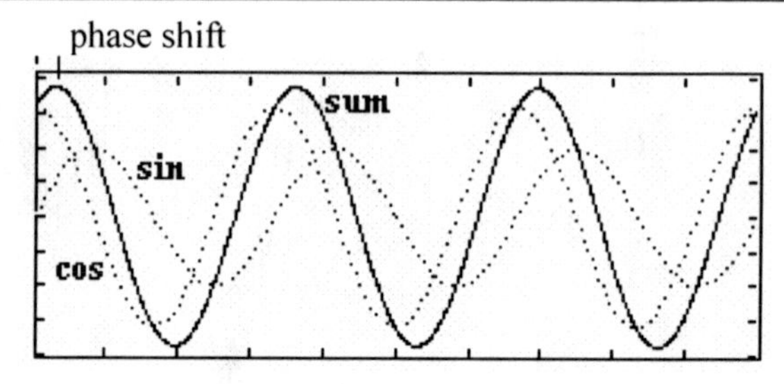


**Figure 3-9. A cosine wave and a sine wave having the same frequency may be summed to obtain a cosine wave with a phase shift.**

Or, looking at it the opposite way, consider a 2 Hz sine wave beginning at zero time, having an amplitude of 6.0 inches and a phase shift of -120 degrees. Figure 3-10 plots this data, first over a four-second time period (upper plot), then a higher resolution plot extending approximately 0.55 seconds. This phase shifted cosine wave may be decomposed into a cosine wave (with no phase shift) and a sine wave. Any phase shifted single frequency periodic function is made of two components: the cosine wave and the sine wave. This relationship between a phase-shifted wave and its cosine and sine wave components is described quantitatively by a trigonometric identity.

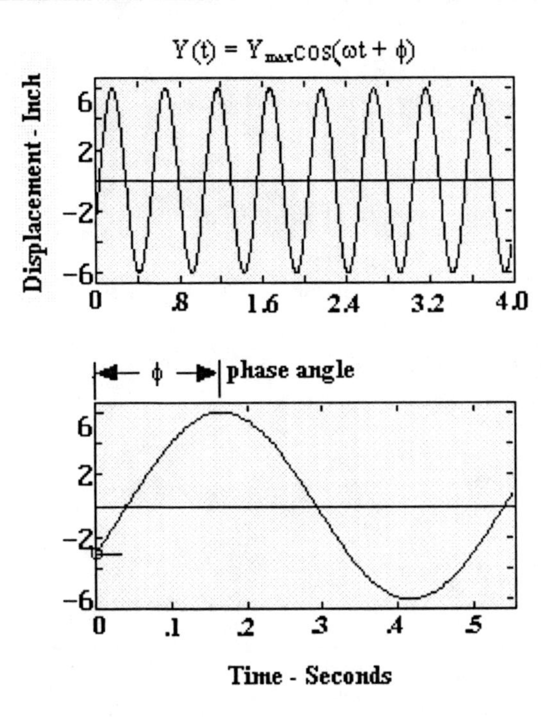


**Figure 3-10. A 2 Hz cosine wave beginning at zero time, having an amplitude of 6.0 inches and a phase shift, $\phi$ = -120 degrees.**

Thus, the handling of phase angle with the Fourier Transform may be understood in the context of the following trigonometric identity for a cosine wave having phase angle, $\phi$:

$$Y_{max}\cos(\omega t + \phi) = A\cos(\omega t) - B\sin(\omega t) \tag{3-1}$$

where

$$A = Y_{max}\cos(\phi) \tag{3-2}$$

and

$$B = Y_{max}\sin(\phi) \tag{3-3}$$

Or, equation (3-1) could be written:

$$Y_{max}\cos(\omega t + \phi) = Y_{max}\cos(\phi)\cos(\omega t) - Y_{max}\sin(\phi)\sin(\omega t) \tag{3-4}$$

So, for the Figure 3-10 cosine and sine wave components we have the amplitudes:

$$A = (6.0)\cos(-120) = -3.0 \text{ Inch}$$

$$B = (6.0)\sin(-120) = -5.20 \text{ Inch}$$

The component cosine and sine waves are plotted along with the phase shifted cosine wave in Figure 3.11.

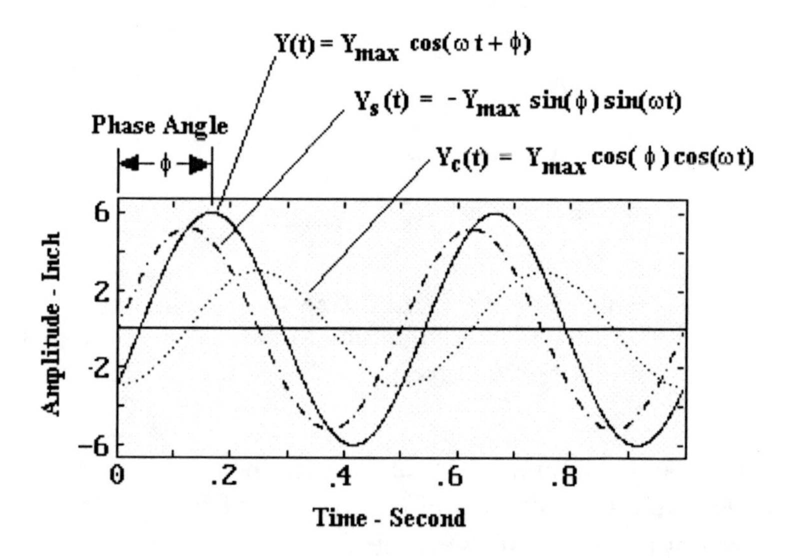


**Figure 3-11. A phase shifted cosine wave has been decomposed into cosine and sine wave components having amplitudes of –3.0 Inches and 5.20 Inches.**

When the Fourier Transform produces frequency domain data from time domain data, it stores the results as the amplitudes of cosine and sine components for each frequency in the spectrum. Generally, an arbitrary time waveform will correspond to a superposition of a very large number of cosine waves of different frequencies, all having different phase angles. The waveform at each frequency could be represented either by a phase shifted amplitude and phase angle or by cosine and sine amplitudes. But, the actual stored data (computer processed data) is in the form of cosine and sine amplitudes. The amplitude relations for the Figure 3-10 example are shown in Figure 3-12.

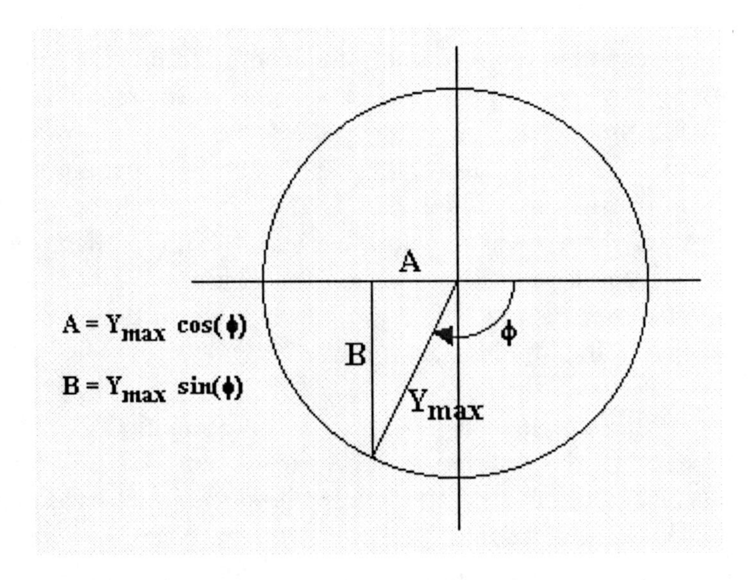


**Figure 3-12. Amplitude relationships between the phase shifted cosine wave of amplitude, $Y_{max}$, and the cosine and sine amplitudes, A and B.**

Recall the –90-degree (-π/2 radians) phase shift for the SDOF displacement response relative to force when sinusoidal excitation is applied at the resonance frequency (Figure 2-3). Recall also the way in which the vibrating mass phase angle shifts with respect to the input base motion throughout a sine sweep process (Figure 2-17).

Figure 2-17 represents a process in which the mass begins vibrating right in step (in phase) with the input base motion at low frequencies, but its motion gradually lags more and more behind the base motion as the frequency sweeps upward. As the input base motion frequency approaches the resonance frequency of the mass and spring, the phase angle moves rapidly through –90 degrees (-π/2 radians), then gradually approaches –180 degrees (-π radians) as the frequency continues sweeping ever higher. We would like to show how the FFT can be used to generate sine sweep plots of this type fairly directly.

If one assumes that the input sine sweep base motion maintains a fixed amplitude of one inch vibration displacement, then a plot of the mass response motion amplitude versus frequency may be obtained simply by performing a FFT on the mass response displacement versus time data. You store all of the response motion data in a computer memory array, then apply the FFT operation to that set of data. The result will be exactly the same as the amp factor plot of Figure 2-17. However, you must utilize both the base motion and the mass response motion to obtain the phase angle versus time. This is because we wish to obtain the phase angle of the mass response with respect to the input base motion.

Now, the simple fact of the FFT representation for the sine sweep process is the following. Performing an FFT of the sine sweep displacement versus time data for both the input base motion and the response mass motion, you can produce both the amp factor versus frequency plot and the phase angle versus frequency plot. That is, if you collect base motion vs. time and mass response motion vs. time, then form the ratio of the FFTs (mass FFT divided by base FFT), you will produce the same amp factor and phase plots as produced by the sine sweep process of Figure 2-17. Figure 3-13 below is a plot of the FFT ratio, mass motion divided by base motion.

Thus, the amplitude and phase is a post-processed form of the direct FFT data. To see how this data is actually processed, we must introduce one more aspect of FFT data. As mentioned previously the FFT performed on a time series directly produces a set of pairs of numbers. One pair of numbers is stored for each frequency in the spectrum, namely the cosine amplitude value and the sine amplitude value. Here's the added feature: Each pair of numbers is taken together to form a complex number, the cosine value representing the real part and the sine value representing the imaginary part. This is a powerful feature of the Fourier data, because it allows the performance of simple but very powerful complex arithmetic.

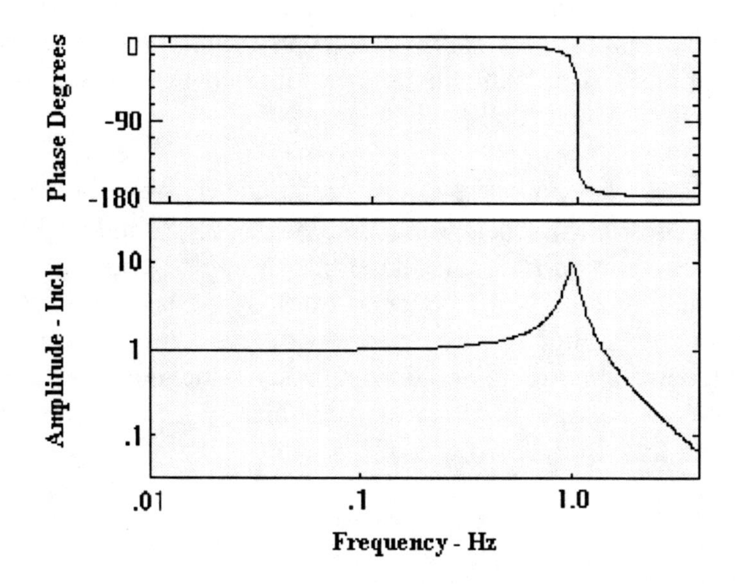


**Figure 3-13. Amplitude and phase angle versus frequency obtained by performing a FFT on the displacement versus time sine sweep data for the vibrating mass of Figure 2-1.**

### 3.5 FFT Computations

Let's look at some of the common computations using FFT data. First, it would be good to have in mind the type of scaling used in the FFT we are working with. Not every computer program implementing an FFT uses the same scaling convention. One program may perform an FFT on a 100 Hz acceleration sine wave of amplitude 1.0 G, producing a frequency plot that exhibits a single line spectrum at 100 Hz, displaying a value of 1.0. This scaling is simply the peak, sine amplitude value. However another program may favor the display of the RMS (Root Mean Square) value, 0.707 $G_{RMS}$. Some programs even display a value of 0.5 G, a number that comes out of a certain straight forward FFT implementation in a natural way. Still others may wish to present a peak-to-peak value of 2.0 $G_{p\text{-}p}$ ("double amplitude").

The scaling convention favored in this text is the RMS value. It is felt that this scaling has more direct physical significance in most applications. Also, mathematical products (power spectra) may be formed in a natural way.

Consider an arbitrary waveform (complicated waveform containing many frequencies) acceleration time history, g(t), and its FFT, G(ω). If the acceleration-time data have been sampled with N data points, then there will be N/2 frequency lines in the spectrum. Remember each frequency line has associated with it a complex number, i.e., a Real (cosine amplitude) and an Imaginary (sine amplitude). Figure 3-14 plots an example acceleration time history along with plots of the Real and Imaginary functions of frequency.

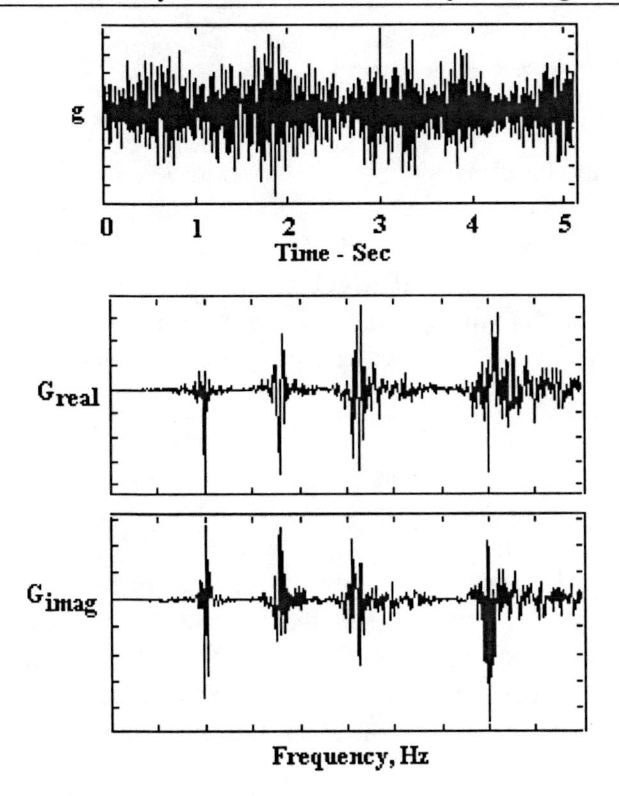


**Figure 3-14. Plots of acceleration-time data and the Real and Imaginary FFT functions of frequency.**

These plots represent the native forms of data as immediately acquired and processed in a computer. Now, we wish to compute absolute magnitude in the frequency domain. That is, we wish to produce a spectrum in which each frequency point displays amplitude scaled as the absolute magnitude in units of $G_{RMS}$. The computation at each ith frequency is:

$$G_{RMS}(\omega_i) = \sqrt{G_{real}^2(\omega_i) + G_{imag}^2(\omega_i)} \qquad (3\text{-}5)$$

However, it is essential to remember that $G_{real}$ and $G_{imag}$ form a complex number. Therefore, the sum of the squares is computed as the complex conjugate product:

$$(G_R + iG_I)(G_R - iG_I) = G_R^2 + G_I^2 \qquad (3\text{-}6)$$

where the letter, i, is the imaginary number, $i = \sqrt{-1}$.

One of the benefits of the RMS scaling is that the overall RMS value (OARMS) of the total acceleration signal may be computed in the frequency domain by taking the square root of the summed squared $G_{RMS}$ values over all N/2 frequencies of the spectrum.

$$G_{OARMS} = \sqrt{\sum_{i=1}^{N/2} G_{RMS}^2(\omega_i)} \qquad (3\text{-}7)$$

At first thought, one might think that the sum of the squares in equation (3-7) should be divided by N/2 to obtain an average. However, the squared values already represent mean squares (averages over the span of each of the N/2 frequency increments), so the summation provides the total mean squared value.

Notice that this is the same overall RMS value that you would obtain if a computation were carried out in the time domain. In particular,

$$G_{OARMS} = \sqrt{\frac{\sum_{i=1}^{N} g^2(t_i)}{N}} = \sqrt{\sum_{i=1}^{N/2} G_{RMS}^2(\omega_i)} \qquad (3\text{-}8)$$

Equation (3-8) is known as Parseval's theorem. In terms of summation of squared terms, one recognizes a conservation of energy principle. Energy (or power) cannot be lost going from the time domain to the frequency domain.

The data of Figure 3-14 is replotted below (Figure 3-15) as $G_{RMS}$ magnitude versus frequency from 0 to 100 Hz. Values are plotted for each of 512 discrete frequency points, giving the appearance of a continuous curve. The $G_{RMS}$ values were obtained using the Real (cosine values) and Imaginary (sine values) of Figure 3-14 and equation 3-5.

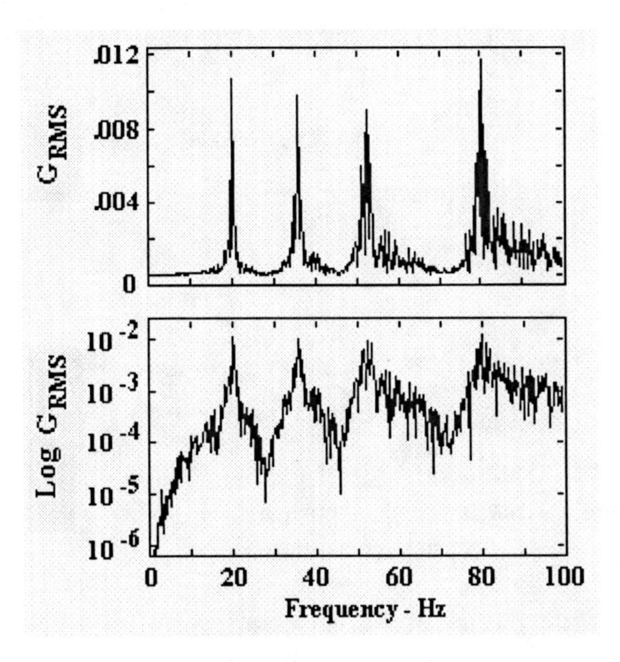


**Figure 3-15. $G_{RMS}$ versus Frequency plotted plotted as linear magnitude (above) and log magnitude (below).**

The spectrum of G-squared values ( $S_{jj}(\omega) = G^2_{RMS}(\omega)$ ) for a single jth acceleration signal is known as an autopower spectrum. It is common to obtain an autopower spectrum that has been averaged over twenty-five or thirty blocks of data (data sets or arrays). For example, consider the acceleration time series of Figure 3-14 to represent one block of data. That time series has a duration of 0.55 seconds and could be just one section of data taken as a snapshot example of a vibration process that continues for an extended period of time. So, the autopower spectrum of that data represents frequency content of a 0.55 second time period.

A more comprehensive representation of the extended vibration process would be obtained if thirty time blocks are captured (16.5 total seconds). An FFT would then be performed for each time block separately to allow computation of the autopower for each time block. The final averaged autopower spectrum, $\overline{S}_{jj}(\omega)$, would be the average of all thirty autopower spectra. The computation would be:

$$\overline{S}_{jj}(\omega) = \overline{G^2_{RMS}}(\omega) = \frac{\sum_{n=1}^{30}\left[(G_R(\omega) + iG_I(\omega))(G_R(\omega) - iG_I(\omega))\right]_n}{30} \quad (3\text{-}9)$$

$$\overline{S}_{jj}(\omega) = \frac{\sum_{n=1}^{30}\left[G^2_R(\omega) + G^2_I(\omega)\right]_n}{30} \quad (3\text{-}10)$$

where $G_R(\omega)$ is the real part and $G_I(\omega)$ is the imaginary part of the FFT of a single time series data block.

Sometimes it is useful to obtain what is called the cross power spectrum. This would be computed for time series data blocks originating from two different accelerometers, for example. Assume we have time series measurements for two separate points on a structure, and we label these points as j and k. The measurements are $g_j(t)$ and $g_k(t)$. The FFTs are represented as $G_j(\omega)$ and $G_k(\omega)$. The averaged cross power spectrum, $\overline{S_{jk}}(\omega)$, would be

$$\overline{S}_{jk}(\omega) = \frac{\sum_{n=1}^{30}\left[\left(G_{jR}(\omega) + iG_{jI}(\omega)\right)\left(G_{kR}(\omega) - iG_{kI}(\omega)\right)\right]_n}{30} \quad (3\text{-}11)$$

The complex product of two FFTs originating from separate signals is often represented as $S_{jk}(\omega) = G_j(\omega) \bullet G^*_k(\omega)$, where the star (asterisk) indicates the FFT complex conjugate. Using this notation, let's look at how easy it is to compute acceleration transmissibility, i.e, the frequency domain ratio of response motion to input base motion (refer to Figures 2-1 and 2-17). We wish to compute the FFT ratio for transmissibility, $T_{jk}(\omega)$.

$$T_{jk}(\omega) = \frac{G_j(\omega)}{G_k(\omega)} \tag{3-12}$$

The trouble here is that both the numerator and the denominator are composed of complex numbers. Therefore, we rationalize the denominator (convert it to real numbers only) by multiplying top and bottom by the complex conjugate of $G_k(\omega)$.

$$T_{jk}(\omega) = \frac{G_j(\omega)}{G_k(\omega)} \bullet \frac{G_k^*(\omega)}{G_k^*(\omega)} \tag{3-13}$$

We see that the numerator is now the cross power spectrum and the denominator is the autopower spectrum of the k measurement (base motion). We immediately recognize a significant advantage of this form of the ratio. It allows us to greatly improve the statistical accuracy of the transmissibility computation by using averaged autopower and cross power spectra:

$$T_{jk}(\omega) = \frac{\overline{S}_{jk}(\omega)}{\overline{S}_{kk}(\omega)} \tag{3-14}$$

Much of the focus of this text will be the development of experimental dynamic modeling of structures using Frequency Response Functions (FRFs). The FRF is mathematically similar to the transmissibility function, except that the input function is an externally applied dynamic force that induces vibration on the structure. The problem is then to obtain an FRF, $h_{jk}(\omega)$, by forming the ratio of a response acceleration FFT to an input force FFT. We use the same formulation as equation (3-13).

$$h_{jk}(\omega) = \frac{G_j(\omega)}{F_k(\omega)} \bullet \frac{F_k^*(\omega)}{F_k^*(\omega)} \tag{3-15}$$

Again the actual computation is accomplished with averaged auto and cross power spectra:

$$h_{jk} = \frac{\bar{S}_{jk}(\omega)}{\bar{S}_{kk}(\omega)} \qquad (3\text{-}16)$$

An example FRF is plotted as Figure 3-16. The response acceleration spectrum used in this example is actually that of Figure 3-15. That response vibration was induced by a random force applied externally to a structure having four resonance frequencies below 100 Hz. That is, Figure 3-16 is the ratio of the Figure 3-15 FFT divided by the FFT of the originally applied force.

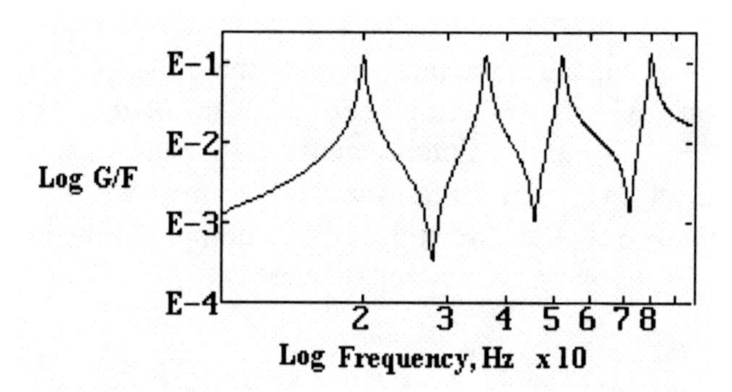


**Figure 3-16. Frequency Response Function (FRF). The ratio of an acceleration FFT divided by an externally applied random force. The force was applied to a structure having four resonance frequencies below 100 Hz (peaks).**

# CHAPTER IV

# COORDINATE REPRESENTATIONS

## 4.1 Generalized Physical Coordinates

So far we have focused on vibration measurement signals and the processing of those signals. Ultimately we must associate those measurements with points on a structure and the vibratory motion of that structure. We do this using coordinate systems and grid points on the structure. We introduce a slightly abstract concept of coordinates and vectors. Consider a representation of the deformation state of a simple beam as a single displacement vector. Figures 4-1 thru 4-3 illustrate the concept. Beam behavior is to be characterized using just three grid points or nodes numbered 1,2 and 3 as shown in Figure 4-1.

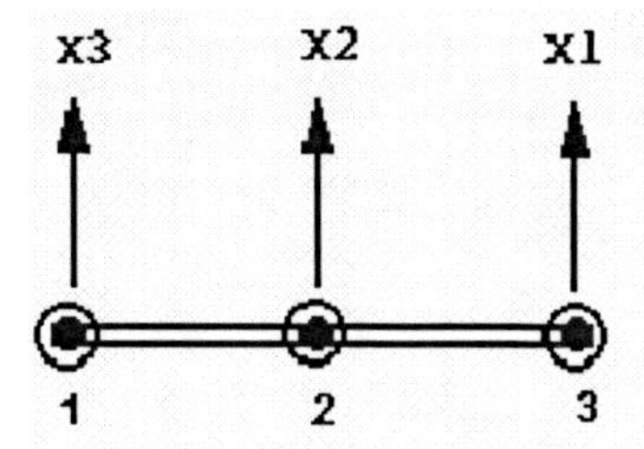


**Figure 4-1. Simple free-free beam. Beam deformation will be characterized by displacements of three node points along their physical coordinates, $X_1$, $X_2$, and $X_3$.**

We assume each node is restrained in all degrees-of-freedom except one (vertical). A generalized coordinate is associated with this free degree-of-freedom, perpendicular to the beam, at each node. So, we have three coordinates, $X_1$, $X_2$ and $X_3$. Figure 4-2 shows an arbitrary deformation state and figure 4-3 represents this deformation with a three dimensional vector. We refer to this representation as a state vector. The components of the vector are of course just the deflections of each of the beam nodes in the direction of their respective generalized coordinates, $X_1$, $X_2$ and $X_3$. The vector, **X,** may be written as:

$$\vec{\mathbf{X}} = X_1 \vec{\mathbf{i}} + X_2 \vec{\mathbf{j}} + X_3 \vec{\mathbf{k}} \tag{4-1}$$

where $\vec{\mathbf{i}}$, $\vec{\mathbf{j}}$ and $\vec{\mathbf{k}}$ are unit vectors. More often vectors in this text will be represented by their components in column matrix form. And for the example, Figure 4-2, the component representation of the vector state is

$$\vec{X} \iff \begin{Bmatrix} X_1 \\ X_2 \\ X_3 \end{Bmatrix} = \begin{Bmatrix} 4.5 \\ -1.5 \\ 2.0 \end{Bmatrix} \tag{4-2}$$

where the symbol <==> indicates correspondence.

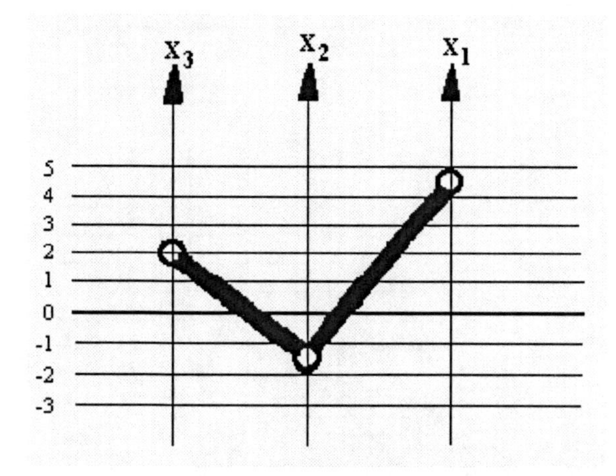


**Figure 4-2 Arbitrary deformation state of free-free beam. Nodes are displaced at coordinates, $X_1$, $X_2$, and $X_3$ with values of 4.5, -1.5 and 2.0 (right to left).**

It is clear that any arbitrary deformation of the three node beam may be represented this way. It is also clear that this concept applies to any structure having an arbitrarily large number of nodes and degrees-of-freedom. It is obviously not possible to draw the state vector for more than three coordinates. But this does not detract from the powerful idea that any arbitrary deformation of the most complicated structure having N degrees-of-freedom is represented by one N-dimensional state vector. Furthermore, the matrix column component correspondence is adequate for the representation of that vector.

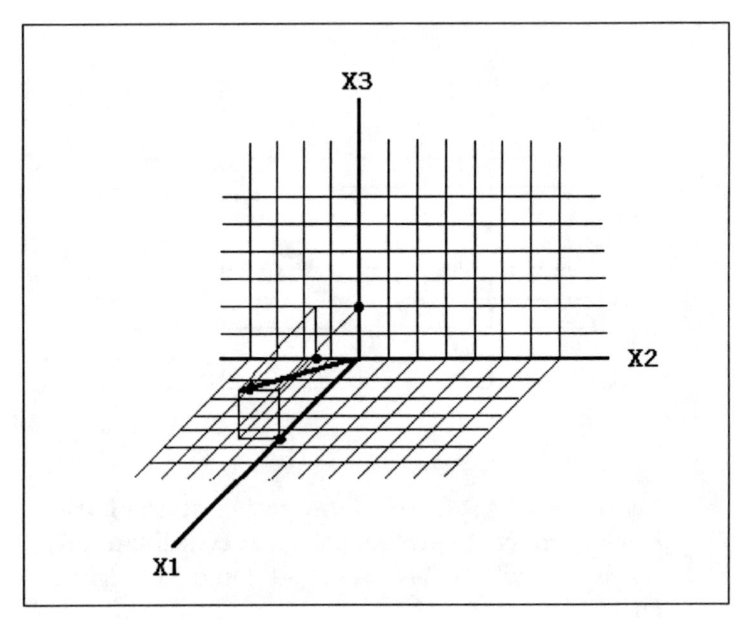


**Figure 4-3. Deformation state of Figure 4-2 represented as a displacement state vector in three dimensions, corresponding to the three generalized coordinates, $X_1$, $X_2$, and $X_3$.**

The column matrix for the three node beam without restraining any of the degrees-of-freedom, i.e., allowing displacements in X, Y and Z directions as well as rotations about each axis at each node, would look like:

$$\vec{\mathbf{X}} \Longleftrightarrow \begin{Bmatrix} X_1 \\ X_2 \\ X_3 \\ X_4 \\ X_5 \\ X_6 \\ X_7 \\ X_8 \\ X_9 \\ . \\ . \\ . \\ . \\ X_{18} \end{Bmatrix} = \begin{Bmatrix} 1X \\ 1Y \\ 1Z \\ 1\Theta_x \\ 1\Theta_y \\ 1\Theta_z \\ 2X \\ 2Y \\ 2Z \\ . \\ . \\ . \\ 3\Theta_z \end{Bmatrix} \tag{4-3}$$

where $\Theta_X$, $\Theta_Y$ and $\Theta_Z$ represent rotations about X, Y and Z directions.

There are other generalized coordinate systems that could be talked about, so to avoid confusion the coordinates discussed up to this point will be refered to as "physical coordinates." It is often possible to directly measure components of a vector in the physical coordinate system, whether it be physical displacements, velocities, accelerations or forces.

## 4.2 Generalized Modal Coordinates

Generalized physical coordinates for a system with a large number of degrees-of-freedom constitute a kind of abstract extrapolation of our normal notions of a three dimensional physical space. Nevertheless, in the case of physical coordinates, it is a fairly easy abstraction to handle, as has been demonstrated in Figures 4-1, 4-2, and 4-3. The concept of modal coordinates is a little more abstract.

Figure 4-4 shows three unique deformation states for the simple three node beam. The deformations are identified as mode shape 1, mode shape 2 and mode shape 3.

Mode shape 1 could be described as a rigid body translation. Mode shape 2 is seen to be
a rigid body rotation, and mode shape 3 is a bending deformation pattern. There is a
certain significance about each of these three special deformation patterns that will become increasingly clear. This signficance makes it worthwhile to identify a special
coordinate system called generalized modal coordinates such that each coordinate in this
system will uniquely associate with one of the special deformation patterns.

The modal coordinate idea is that a single abstract generalized coordinate, using a coordinate system different than generalized physical coordinates, may be used to describe the beam deformation when it is in a deformation state corresponding to one of the unique mode shapes. The "modal coordinates" used to define the amount of displacement (amplitude of displacement) for each of the three mode shapes are $\underline{X}_1$, $\underline{X}_2$ and $\underline{X}_3$ (underlined characters will be used throughout the text to represent modal coordinates).

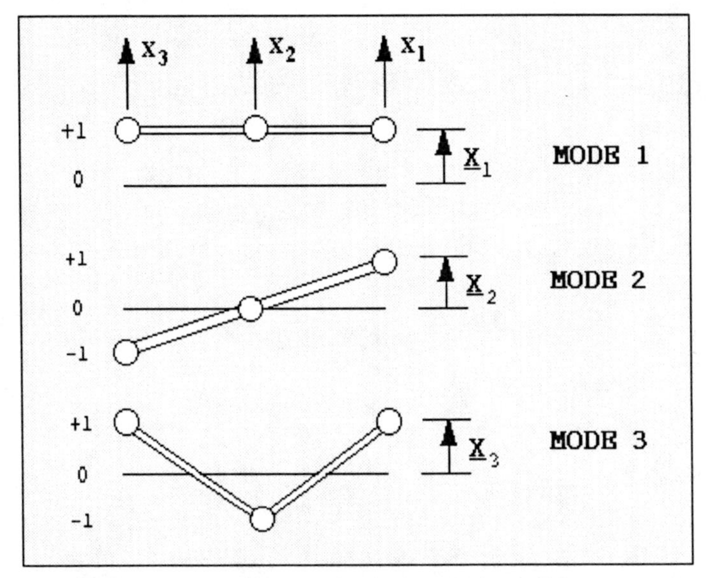


**Figure 4-4. Three unique deformation patterns corresponding to abstract generalized "modal" coordinates, $\underline{X}_1$, $\underline{X}_2$, and $\underline{X}_3$.**

Later, the dynamics associated with mode shapes will be discussed. For now, attention is to be focused on the ease with which the most arbitrary deformation states can be understood in modal coordinates. Figures 4-5, 4-6 and 4-7 show the displacement vector states for each of the three mode shapes. Remember that a vector... is a vector... is a vector, and it doesn't matter what frame of reference is being used to draw the picture of the vector. The vector can be drawn in either the physical coordinate frame of reference or the modal coordinate frame of reference. The point is that the vectors are the same:

$$\vec{\mathbf{x}}_A = \vec{\underline{\mathbf{X}}}_A \qquad (4\text{-}4)$$

$$\vec{\mathbf{x}}_B = \vec{\underline{\mathbf{X}}}_B \qquad (4\text{-}5)$$

$$\vec{\mathbf{x}}_C = \vec{\underline{\mathbf{X}}}_C \tag{4-6}$$

Each of the modal deformation state vectors, Figures 4-5, 4-6 and 4-7, can be recognized as basis vectors for the modal coordinate system, $\underline{X}_1$, $\underline{X}_2$ and $\underline{X}_3$. The modal coordinate system is shown in Figure 4-8.

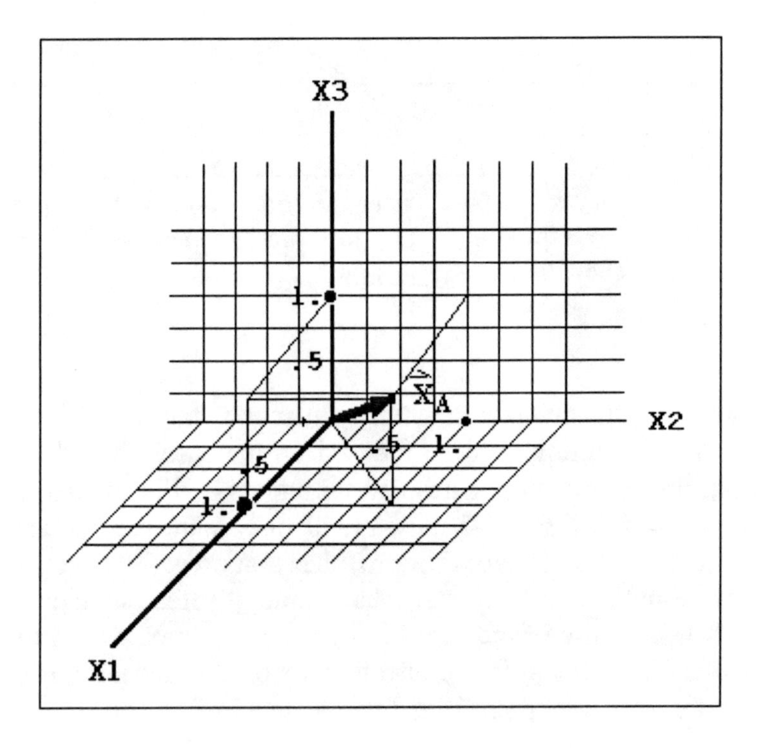


**Figure 4-5. State vector for mode shape number 1. The vector points in the direction of modal coordinate, $\underline{X}_1$. A base vector for the first modal coordinate may be defined with this vector.**

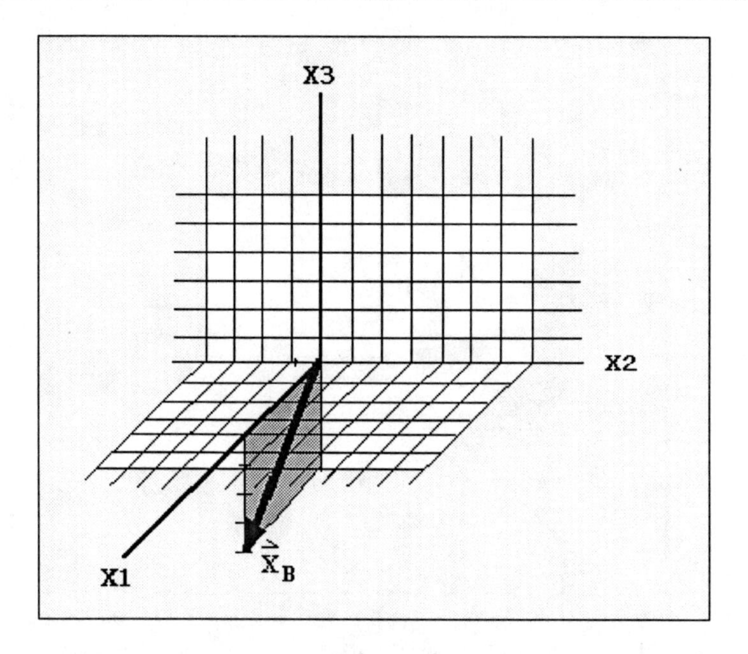


**Figure 4-6. State vector for deformation corresponding to mode shape number 2. This vector points in the direction of modal coordinate $\underline{X}_2$.**

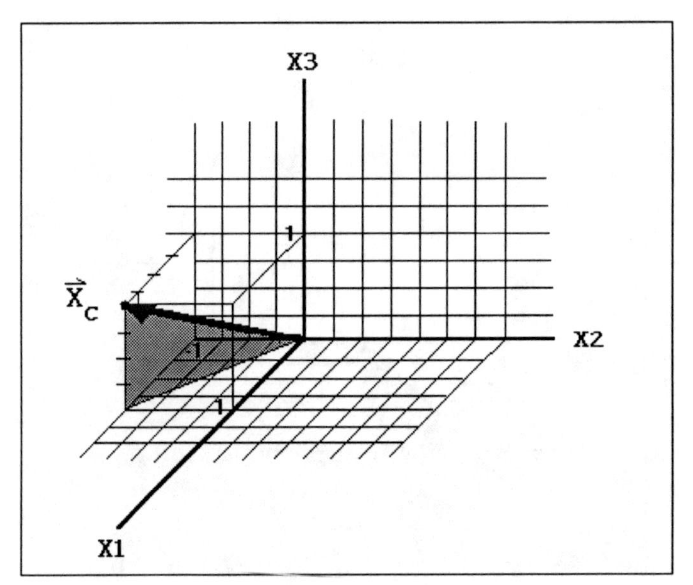


**Figure 4-7. State vector for mode shape number 3. The vector points in the direction of modal coordinate, $\underline{X}_3$.**

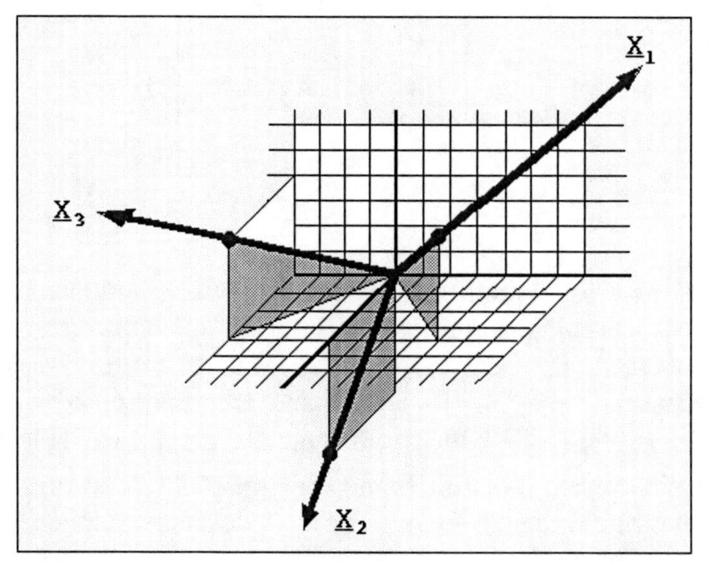


**Figure 4-8. Generalized modal coordinates. Displacement along one coordinate, say $\underline{X}_1$, corresponds to a deformation of mode shape number one, with deformation amplitude proportional to distance along the coordinate axis.**

The modal base vectors are shown together in Figure 4-9. The column matrices corresponding to modal coordinate system components for each of the base vectors are the following.

$$\vec{\mathbf{X}}_A = \underline{\vec{\mathbf{b}}}_1 \Leftrightarrow \begin{Bmatrix} \underline{b}_1 \\ 0 \\ 0 \end{Bmatrix} \tag{4-7}$$

$$\vec{\mathbf{X}}_B = \underline{\vec{\mathbf{b}}}_2 \Leftrightarrow \begin{Bmatrix} 0 \\ \underline{b}_2 \\ 0 \end{Bmatrix} \tag{4-8}$$

$$\vec{\mathbf{x}}_C = \vec{\underline{\mathbf{b}}}_3 \Leftrightarrow \begin{Bmatrix} 0 \\ 0 \\ \underline{b}_3 \end{Bmatrix} \tag{4-9}$$

These same base vectors represented by the component matrices in physical coordinates are given below. The physical coordinate components for each modal base vector are given the special designation, "mode coefficients", and are represented by the greek letter, $\Psi_{\mathbf{jr}}$ (psi), which will typically carry two indices, j and r. The inside index, j, identifies the node number (physical location) and the outer right index, r, identifies the mode shape number.

$$\vec{\mathbf{x}}_A = \vec{b}_1 \Leftrightarrow \begin{Bmatrix} \psi_{11} \\ \psi_{21} \\ \psi_{31} \end{Bmatrix} = \begin{Bmatrix} 1 \\ 1 \\ -1 \end{Bmatrix} \tag{4-10}$$

$$\vec{\mathbf{x}}_B = \vec{b}_2 \Leftrightarrow \begin{Bmatrix} \psi_{12} \\ \psi_{22} \\ \psi_{32} \end{Bmatrix} = \begin{Bmatrix} 1 \\ 0 \\ -1 \end{Bmatrix} \tag{4-11}$$

$$\vec{\mathbf{x}}_C = \vec{b}_3 \Leftrightarrow \begin{Bmatrix} \psi_{13} \\ \psi_{23} \\ \psi_{33} \end{Bmatrix} = \begin{Bmatrix} 1 \\ -1 \\ 1 \end{Bmatrix} \tag{4-12}$$

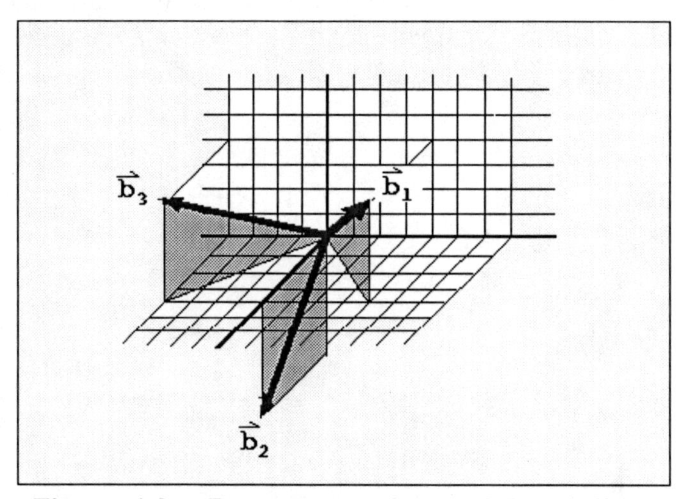


**Figure 4-9. Base vectors for modal coordinates. Each base vector has three physical coordinate components that define a mode shape pattern.**

The complete set of mode shapes can be represented in one square matrix where columns 1, 2 and 3 of the matrix are just the column matrices of mode coefficients given above. It is seen below that each column of the matrix is a mode shape and corresponds to a modal coordinate degree-of-freedom, whereas each row of the matrix corresponds to a degree-of-freedom in physical coordnates.

$$\begin{bmatrix} & & \\ & \Psi & \\ & & \end{bmatrix} = \begin{bmatrix} \Psi_{11} & \Psi_{12} & \Psi_{13} \\ \Psi_{21} & \Psi_{22} & \Psi_{23} \\ \Psi_{31} & \Psi_{32} & \Psi_{33} \end{bmatrix} = \begin{bmatrix} 1 & 1 & 1 \\ 1 & 0 & -1 \\ 1 & -1 & 1 \end{bmatrix} \qquad (4\text{-}13)$$

## 4.3 Mode Shape Superposition

Figure 4-10 shows another arbitrary deformation state of the three node beam (different from Figure 4-2). The state vector, $\vec{X}_D$, is shown in Figure 4-11 using the physical coordinate frame of reference. The physical coordinate components are given below as a column matrix .

$$\vec{X}_D \Leftrightarrow \begin{Bmatrix} x_1 \\ x_2 \\ x_3 \end{Bmatrix} = \begin{Bmatrix} 2.0 \\ 3.0 \\ -4.0 \end{Bmatrix} \qquad (4\text{-}14)$$

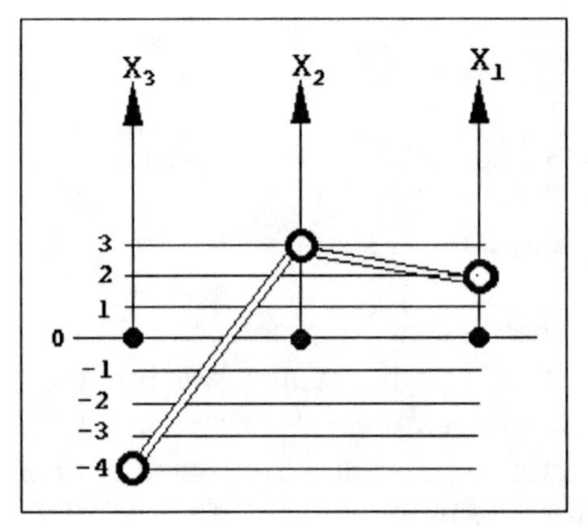


**Figure 4-10. An arbitrary example deformation state of the three-node free-free beam. The node deflections are: X1 = 2.0, X2 = 3.0, and X3 = -4.0**

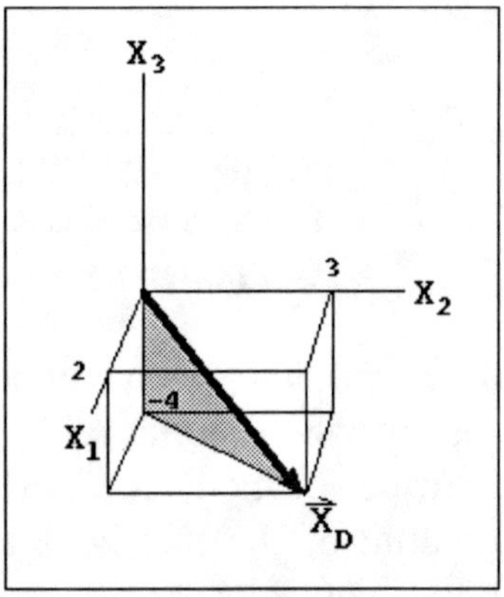

**Figure 4-11. State vector, $\vec{X}_D$, for the arbitrary deformation of Figure 4-10.**

This same state vector is shown again in the modal coordinate frame of reference, Figure 4-12. The modal components in column matrix form are

$$\underline{\vec{X}}_D \Leftrightarrow \begin{Bmatrix} \underline{x}_1 \\ \underline{x}_2 \\ \underline{x}_3 \end{Bmatrix} = \begin{Bmatrix} 1.0 \\ 3.0 \\ -2.0 \end{Bmatrix} \qquad (4\text{-}15)$$

Now, the original deformation state shown in Figure 4-10 can be recognized as a linear superposition of the mode shapes

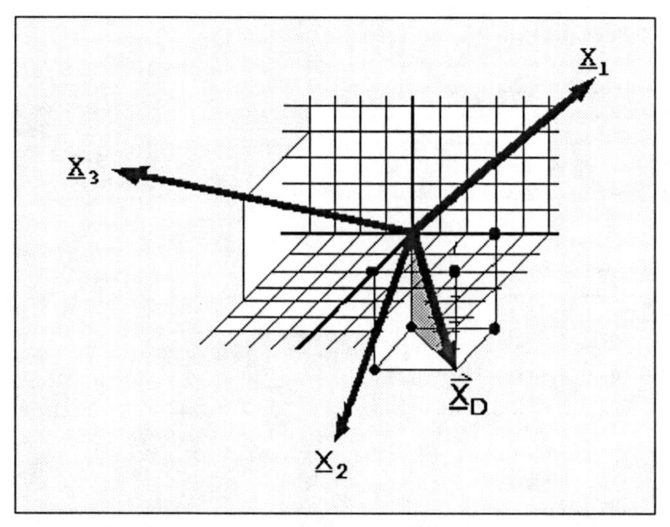


**Figure 4-12. Same state vector. This time it is shown in the modal coordinate system. The magnitude and direction of the vector are the same, but the physical coordinate components (marked with dots) are of course not the same as the modal coordinate components.**

having modal coordinate displacement amplitudes of 1.0, 3.0 and -2.0 for modal coordinates, $\underline{X}_1$, $\underline{X}_2$ and $\underline{X}_3$. The three mode shape deformations to be superimposed are shown in Figure 4-13. Figure 4-14 shows the superposition process as summed modal vectors producing the expected final state vector.

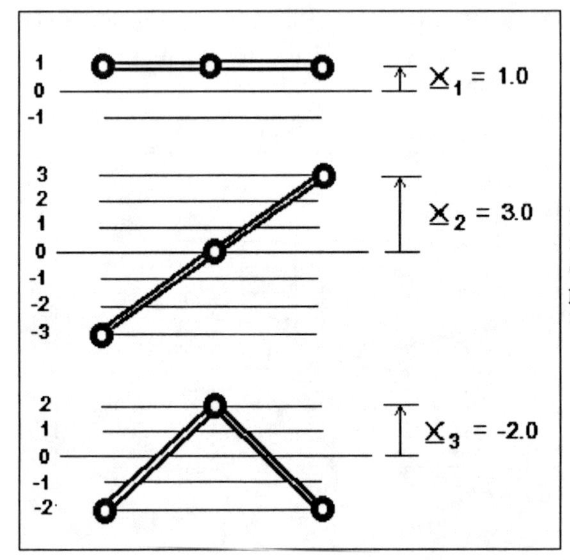


**Figure 4-13. The deformation of Figure 4-10 can be developed from a superposition of these modal deformations.**

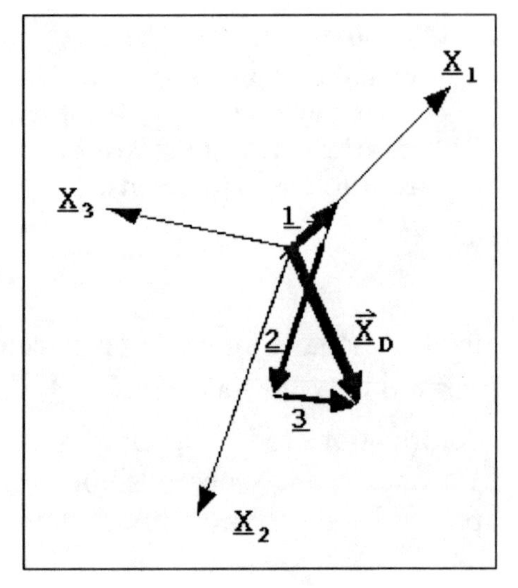


**Figure 4-14. Modal deformations of Figure 4-13 superimposed by vector summation. The final deformation state is represented by the displacement state vector, $\vec{\underline{X}}_D$, shown here in the modal coordinate reference frame. Each mode deformation of Figure 4-13 is represented by a vector component directed along the corresponding modal coordinate.**

## 4.4 Coordinate Transformations

Section 4.3 showed that an arbitrary deformation state represented by physical degree-of-freedom node displacements may be decomposed into mode shape components. Conversely, individual mode shapes having suitable amplitudes may be summed in vector fashion to produce some desired state of physical degree-of-freedom deformation of a structure. Of course there must be a coordinate transformation that relates physical components of a state vector to the corresponding modal coordinate components. The matrix of mode coefficients, equation (4-13), provides just that transformation matrix. The transformation from modal coordinates to physical coordinates is

$$\begin{Bmatrix} X_1 \\ X_2 \\ X_3 \end{Bmatrix} = \begin{bmatrix} \psi_{11} & \psi_{12} & \psi_{13} \\ \psi_{21} & \psi_{22} & \psi_{23} \\ \psi_{31} & \psi_{32} & \psi_{33} \end{bmatrix} \begin{Bmatrix} \underline{X}_1 \\ \underline{X}_2 \\ \underline{X}_3 \end{Bmatrix} \qquad (4\text{-}16)$$

Or, in compact form,

$$\begin{Bmatrix} X \end{Bmatrix} = \begin{bmatrix} \Psi \end{bmatrix} \begin{Bmatrix} \underline{X} \end{Bmatrix} \qquad (4\text{-}17)$$

The modal coordinates may of course be found, given the physical coordinate deformations, by using the inverse of the mode shape matrix.

$$\{\underline{X}\} = [\Psi]^{-1}\{X\} \qquad (4\text{-}18)$$

Let's check out the mode shape matrix for the three-node beam to see if it will perform properly as a transformation matrix. Using the modal deformation amplitudes, $\{\underline{X}\}$, depicted by the modal components of equation (4-15), the corresponding physical coordinate components, $\{X\}$, are computed as

$$\begin{Bmatrix} x_1 \\ x_2 \\ x_3 \end{Bmatrix} = \begin{Bmatrix} 2.0 \\ 3.0 \\ -4.0 \end{Bmatrix} = \begin{bmatrix} 1 & 1 & 1 \\ 1 & 0 & -1 \\ 1 & -1 & 1 \end{bmatrix} \begin{Bmatrix} 1.0 \\ 3.0 \\ -2.0 \end{Bmatrix} \tag{4-19}$$

And it can be seen that the mode shape matrix does correctly play the role of a coordinate transformation matrix. To verify that the modal coordinate components are correctly computed, given state vector components in physical coordinates, first invert the transformation matrix to get

$$[\Psi]^{-1} = \begin{bmatrix} 0.25 & 0.5 & 0.25 \\ 0.5 & 0.0 & -0.5 \\ 0.25 & -0.5 & 0.25 \end{bmatrix} \tag{4-20}$$

Using the physical coordinate components of the state vector as listed in (4-14), the modal components are correctly computed as

$$\begin{Bmatrix} \underline{x}_1 \\ \underline{x}_2 \\ \underline{x}_3 \end{Bmatrix} = \begin{Bmatrix} 1.0 \\ 3.0 \\ -2.0 \end{Bmatrix} = \begin{bmatrix} 0.25 & 0.5 & 0.25 \\ 0.5 & 0.0 & -0.5 \\ 0.25 & -0.5 & 0.25 \end{bmatrix} \begin{Bmatrix} 2 \\ 3 \\ -4 \end{Bmatrix} \tag{4-21}$$

It should go without saying: Once you have the coordinate transformation between physical coordinates and modal coordinates, you then have the transformation between physical coordinates for components of *any* vector. Indeed, this actually provides the definition of a vector. The definition of a vector in elementary courses has always referred to a quantity having "magnitude and direction." A more sophisticated definition of a vector is based on the criteria that components must transform from one coordinate system to another using the same transformation as the coordinates. For example, the components, $\{F\}$, represent a vector force, by definition, only if these components transform from modal coordinates to physical coordinates using the same transformation as (1-17), $\{F\}=[\Psi]\{\underline{F}\}$. And the components, $\{A\}$, represent vector acceleration only if these components transform in that same manner, $\{A\}=[\Psi]\{\underline{A}\}$. This same transformation is a requirement for *any* vector. Otherwise it simply is not a vector.

## 4.5 Two D.O.F. Mode Shapes

The following example uses just two degrees-of-freedom in order to make the analysis of vector components easy to visualize.

A system is comprised of two masses and two springs, depicted in Figure 4-15 with physical coordinates, $X_1$ and $X_2$ associated with the motion of mass 1 and mass 2.

The system will have two resonance frequencies, $\nu_1$ and $\nu_2$. It will be assumed that the system will vibrate naturally at the first frequency, $\nu_1$ , maintaining oscillatory motion in the deformation pattern of Figure 4-16. Similarly the system will vibrate at the second frequency, $\nu_2$ , maintaining the pattern of Figure 4-17.

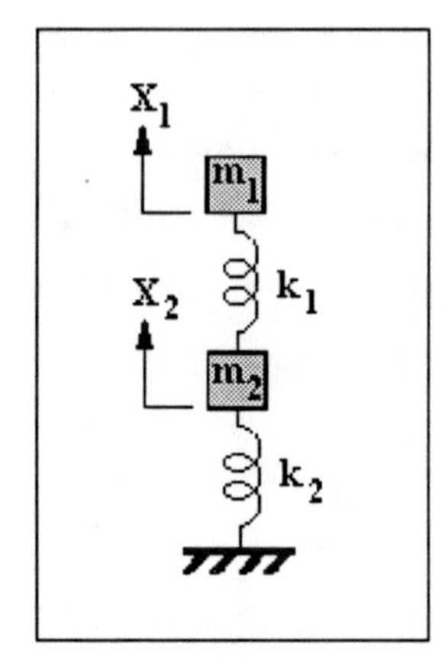


**Figure 4-15. Two DOF spring-mass system with physical coordinates, $X_1$ and $X_2$.**

Figures 4-16 and 4-17 describe mode shapes 1 and 2. These deformation states correspond to a snapshot photograph of system sinusoidal vibration at each frequency at just the instant maximum deflection is reached. The solution to the two degree-of-freedom system will be derived in detail later, but for now the solution is accepted as represented by mode shapes and resonance frequencies of Figures 16 and 17.

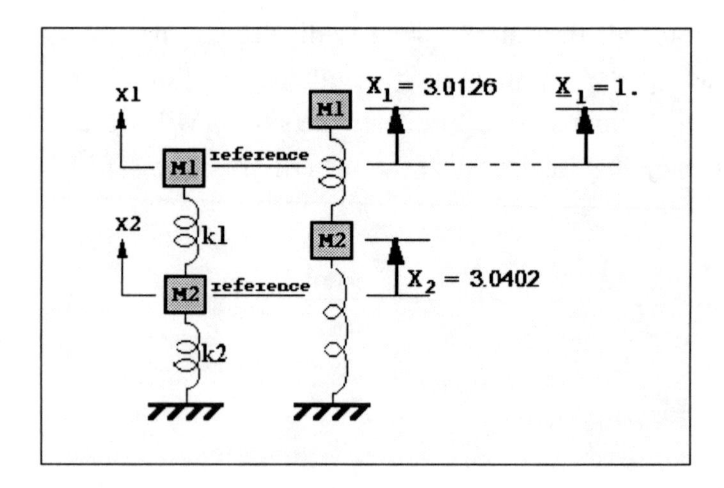


**Figure 4-16. Two DOF mode shape number 1. The system vibrates naturally in this deformation pattern at the frequency, $\nu_1$ = 4.799 Hz.**

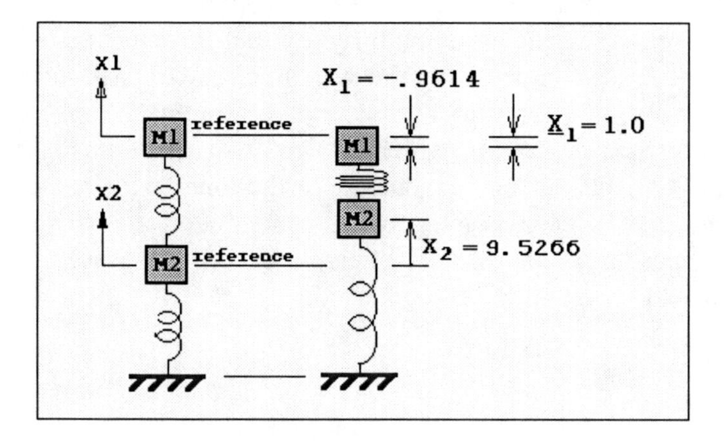


**Figure 4-17. Two DOF mode shape number 2. Resonance frequency, $\nu_2$ = 52.81 Hz.**

We have the usual relationship between deformation components of the physical coordinate system and the modal coordinate system,

$$\begin{Bmatrix} X_1 \\ X_2 \end{Bmatrix} = \begin{bmatrix} \psi_{11} & \psi_{12} \\ \psi_{21} & \psi_{22} \end{bmatrix} \begin{Bmatrix} \underline{X}_1 \\ \underline{X}_2 \end{Bmatrix} \tag{4-22}$$

And using the deformation state components of Figures 4-16 and 4-17 to define the mode shape base vectors, the transformation matrix has the mode coefficients

$$\begin{bmatrix} \psi_{11} & \psi_{12} \\ \psi_{21} & \psi_{22} \end{bmatrix} = \begin{bmatrix} 3.0126 & -0.9614 \\ 3.0402 & 9.5266 \end{bmatrix} \tag{4-23}$$

The coordinate systems for the two degree-of-freedom system are shown in Figure 4-18. Right away it is noticed that the modal coordinates are not at all orthogonal (perpendicular). Actually, this is the case for the vast majority of real structures. This could be a source of confusion to the dynamicist who has always heard that "modes of vibration are always orthogonal", i.e., normal modes. Later, the context in which modes are regarded as "orthogonal" will be clarified, using the phrase "orthogonal with respect to the mass matrix." But, it is important to understand that the geometric mode shapes are not at all orthogonal. This will be proven later, but first it would be useful to review the concept of scalar invariants.

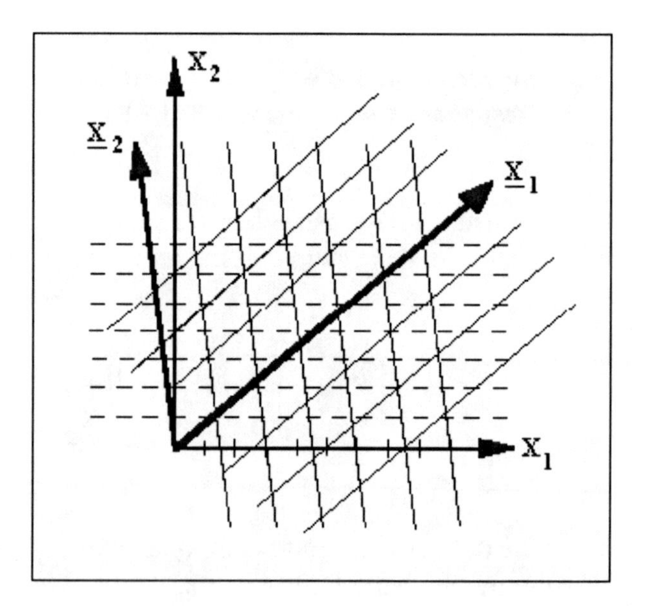


**Figure 4-18. Physical coordinates (X1 and X2) and modal coordinates ($\underline{X1}$ and $\underline{X2}$) for the two DOF of Figures 4-15, 4-16, and 4-17. The physical coordinates are orthogonal, but notice that modal coordinates are not perpendicular to each other.**

## 4.6 Scalar Invariants

The way of computing the scalar product of two vectors in modal coordinates as compared to the computation of that same scalar product in physical coordinates leads to some important insight into aspects of modal analysis.

Consider the process in which two sinusoidal oscillatory forces, $f_1$ and $f_2$, are applied to our two degree-of-freedom system of Figure 4-15. Assume that both forces are applied at the same frequency, and that frequency is not at either resonance, rather some frequency between the first and second resonance frequencies. Furthermore, the force magnitudes are not adjusted to favor the excitation of one modal deformation pattern over the other. The effect will be to drive the system into an oscillatory deformation pattern that is a superposition of the two mode shapes. Figure 4-19 is a photograph snapshot of the system at just the instant maximum deflection is reached. Figure 4-20 shows the state vector, **R**, of the system at that instant.

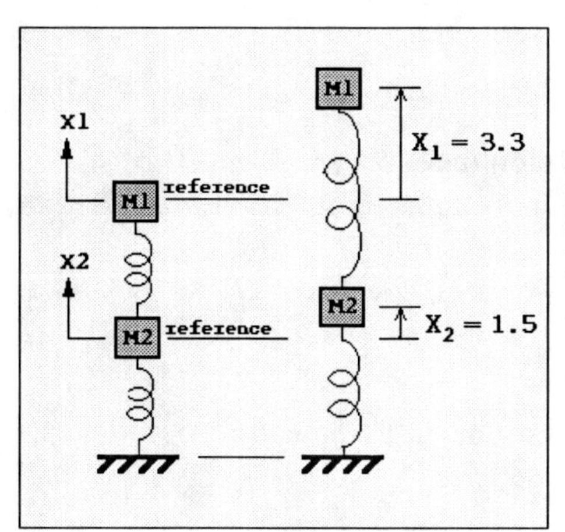


**Figure 4-19. Snapshot of two DOF deformation state at instant of peak displacement. The magnitude of the state vector for this deformation is computed by using the Pythagorean theorem:**

$$|M| = \sqrt{(3.3)^2 + (1.5)^2}$$

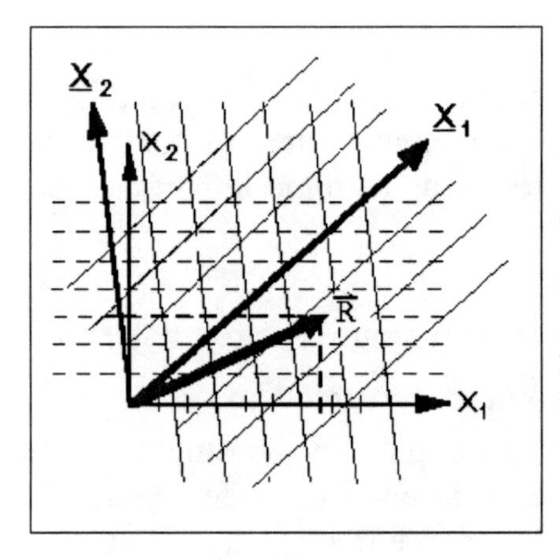


**Figure 4-20. State vector, R , represents the deformation state of the two DOF system of Figure 4-19. It can be seen that the magnitude can be computed using the Pythagorean theorem in physical coordinates. However, this will not work if you try to use modal components, since modal coordinates are not perpendicular.**

Now, we wish to compute the magnitude of the state vector in each coordinate system, and we expect that the magnitudes will be exactly the same. The state vector is exactly the same vector, regardless of the frame of reference used, and the magnitudes should be the same, i.e.,

$$\vec{\mathbf{R}} = \underline{\vec{\mathbf{R}}} \tag{4-24}$$

And the scalar product (magnitude squared) must be the same, independent of the coordinate system. This is referred to as scalar invariance:

$$\vec{\mathbf{R}} \bullet \vec{\mathbf{R}} = \underline{\vec{\mathbf{R}}} \bullet \underline{\vec{\mathbf{R}}} \tag{4-25}$$

The vector dot product is computed in physical coordinates using matrix notation as follows.

$$\vec{\mathbf{R}} \bullet \vec{\mathbf{R}} = \begin{Bmatrix} R \end{Bmatrix}^T \begin{Bmatrix} R \end{Bmatrix} = R_1^{\ 2} + R_2^{\ 2} \qquad (4\text{-}26)$$

This works fine for physical coordinates, because they are orthogonal coordinates and you just use the Pythagorean theorem. However, the Pythagorean theorem just doesn't apply to modal coordinates, i.e., the magnitude squared is not equal to the sum of the products of the modal components:

$$\underline{\vec{\mathbf{R}}} \bullet \underline{\vec{\mathbf{R}}} \neq \begin{Bmatrix} \underline{R} \end{Bmatrix}^T \begin{Bmatrix} \underline{R} \end{Bmatrix} \qquad (4\text{-}27)$$

The solution to this problem may seem too involved at first, but it is well worth pursuing. The mathematics of scalar products for coordinates other than cartesian orthogonal is formalized in text books on tensor analysis. Our modal coordinates are sometimes referred to as affine. Tensor analysis concerns itself with allowable transformations under the centered affine group. The approach to the scalar product problem is actually quite simple. Just insert an appropriate matrix into the matrix multiplication such that the scalar product in modal coordinates is exactly the same as in physical coordinates:

$$\{X\}^T\{X\} = \{\underline{X}\}^T[G]^T\{\underline{X}\} \qquad (4\text{-}28)$$

The matrix, $[G]^T$, is known as a metric matrix. If the transposed modal column vector is grouped together with the metric matrix, you can see that another set of coordinates could be inferred:

$$\{\underline{X}'\}^T = \{\underline{X}\}^T[G]^T \qquad (4\text{-}29)$$

Using these new coordinates together with modal coordinates allows scalar products to be computed in the same matrix form as physical coordinates:

$$\{X\}^T\{X\} = \{\underline{X}'\}^T\{\underline{X}\} \tag{4-30}$$

So, a new coordinate system is introduced which will be a companion to the modal coordinate system. It will be called the *contravariant* modal coordinate system. The modal coordinates we have been using are actually known in tensor analysis as *covariant* coordinates. The contravariant modal coordinates are selected so as to have a specific relation to the modal coordinates (the term, modal coordinates, will always imply covariant modal coordinates).

Figure 4-21 shows the contravariant modal coordinates along with physical and modal coordinates. The key to the relation between covariant modal coordinates and contravariant modal coordinates is that each covariant coordinate is perpendicular to all but one of the contravariant coordinates. This is what allows the Pythagorean theorem to work when using contravariant and covariant coordinates in combination.

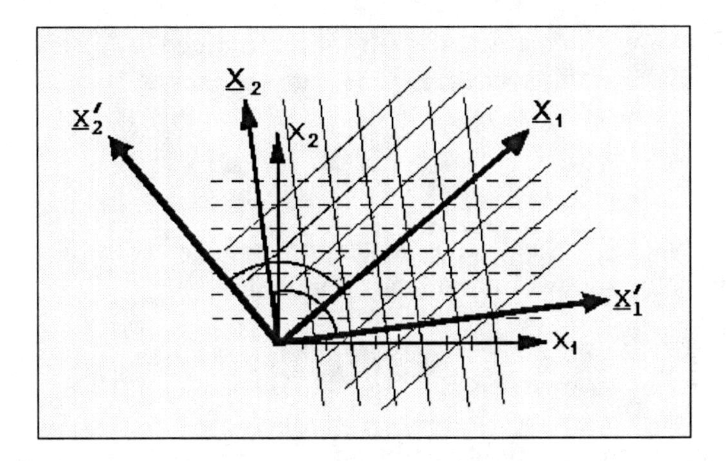


**Figure 4-21. Contravariant modal coordinates, $\underline{X}'_1$ and $\underline{X}'_2$ , are included with covariant modal coordinates and physical coordinates. Contravariant modal coordinate, $\underline{X}'_2$ , is perpendicular to covariant modal coordinate, $\underline{X}_1$. Pythagorean theorem works for computing magnitude in modal coordinates if you match up contravariant with covariant for products.**

Notice now the very signficant implication of forming the scalar product for work, i.e., vector force times vector displacement. Obviously, if energy is to be conserved, this product must be the same in physical coordinates as in modal coordinates:

$$\{F\}^T\{X\} = \{\underline{F}'\}^T\{\underline{X}\} \tag{4-31}$$

The important physical implication here is that to produce a deflection that is a pure mode shape, you must apply a single contravariant modal force. In the language of coordinates, a force vector applied along the contravariant modal coordinate number 1 produces a deflection along covariant coordinate number 1. The contravariant force is also referred to in some literature as a "left vector" (a row vector multiplying from the left).

But how can the metric matrix be computed? To see this, start again with the scalar product in physical coordinates.

$$\{X\}^T\{X\} = \{X\}^T\{X\} \tag{4-32}$$

Replace physical coordinates on the right hand side with their transformations from modal coordinates.

$$\{X\}^T\{X\} = \{\underline{X}\}^T[\Psi]^T[\Psi]\{\underline{X}\} \tag{4-33}$$

Comparing the right hand side of (4-33) to the right hand side of (4-28), we find by inspection that the metric matrix, $[G]^T$ , is the product, $[\Psi]^T[\Psi]$.

$$[G]^T = [\Psi]^T[\Psi] \tag{4-34}$$

Notice that $[G]$ is symmetric, i.e.,

$$[G]^T = [G] \tag{4-36}$$

Or,

$$\left[\Psi^T\Psi\right]^T = [\Psi]^T[\Psi] \tag{4-37}$$

Transposing both sides of (4-29),

$$\{\underline{X}'\} = [G]\{\underline{X}\} \tag{4-37}$$

Using (4-34) to substitute for [G] gives the transformation from covariant modal coordinates, $\{\underline{X}\}$, to contravariant modal coordinates, using the product, $[\Psi]^T[\Psi]$.

$$\{\underline{X}'\} = [\Psi]^T[\Psi]\{\underline{X}\} \tag{4-38}$$

## 4.7 Transformation From Physical Coordinates To Contravariant Modal Coordinates

The transformation from physical coordinates to contravariant coordinates has frequently used applications. In the test lab it is sometimes desireable to force an oscillatory deformation pattern on a dynamic structure such that it vibrates in just one pure mode shape, $\{\Psi\}_r$, corresponding to the covariant modal displacement vector, $\underline{\mathbf{X}}_r$. The correct set of physical forces for accomplishing this would be those corresponding to the contravariant vector force, $\underline{\mathbf{F}}_r'$. Typically, one would like to know the coordinate transformation so that the desired physical forces (applied using a set of electrodynamic shakers) may be established for a given contravariant modal force component, $\{\underline{F}_r'\}$.

Based on (4-38), we can easily see the transformation from physical to contravariant modal coordinates. Since the product, $[\Psi]\{\underline{X}\}$, on the right side of (4-38) is just the transformation from modal to physical coordinates, i.e., equation (4-17), it may be replaced with physical displacement, $\{X\}$.

$$\left\{\underline{X'}\right\} = \left[\quad \Psi \quad\right]^T \left\{X\right\} \tag{4-39}$$

Force vector components must transform the same way:

$$\left\{\underline{F'}\right\} = \left[\quad \Psi \quad\right]^T \left\{F\right\} \tag{4-40}$$

The physical forces may be obtained from the contravariant modal forces by

$$\left\{F\right\} = \left[\Psi\right]^{-T} \left\{\underline{F}'\right\} \tag{4-41}$$

The set of physical forces needed to excite mode shape number 3, only, for example, would be computed by setting all components of $\{\underline{F}'\}$ to zero (right side of (4-41), except for $\underline{F}_3'$. Then, using (4-41),

$$\{F\}_{(\text{mod e3})} = [\Psi]^{-T} \begin{Bmatrix} 0 \\ 0 \\ \underline{f}_3' \\ 0 \\ \vdots \\ 0 \end{Bmatrix} \tag{4-42}$$

Sometimes, the transformation matrix, $[\Psi]^{-T}$, of (4-41) is not available to the engineer, or he may not be aware of its importance in a given case. So, it is not unusual to drive the structure using physical forces in proportion to the mode shape deformation itself. The error in achieving contravariant modal force purity can be assessed by putting in actual physical forces used into the right side of (4-40) and looking at the size of the undesired contravariant force components:

$$\begin{Bmatrix} \underline{f}_1' \\ \underline{f}_2' \\ \underline{f}_3' \\ \vdots \\ \underline{f}_n' \end{Bmatrix} = [\Psi]^T \begin{Bmatrix} f_1 \\ f_2 \\ f_3 \\ \vdots \\ f_n \end{Bmatrix} \tag{4-43}$$

Although $[\Psi]^T$ of (4-39) does the right job getting from physical coordinates to contravariant modal coordinates, the set of base vectors, $[\Psi']$, that directly represent the transformation from contravariant modal coordinates to physical coordinates should be shown. There must be such a matrix, $[\Psi']$, so that

$$\begin{Bmatrix} X \end{Bmatrix} = \begin{bmatrix} & \Psi' & \end{bmatrix} \begin{Bmatrix} \underline{X'} \end{Bmatrix} \tag{1-44}$$

and

$$\begin{Bmatrix} \underline{X'} \end{Bmatrix} = \begin{bmatrix} & \Psi' & \end{bmatrix}^{-1} \begin{Bmatrix} X \end{Bmatrix} \tag{1-45}$$

Comparing (1-45) to (1-39),

$$[\Psi']^{-1} = [\Psi]^{T} \tag{4-46}$$

and

$$[\Psi']^{T} = \left[\Psi^{-1}\right] \tag{4-47}$$

This is consistent with orthogonality of the covariant and contravariant mode shapes, i.e.,

$$\begin{bmatrix} & & \\ & \Psi' & \\ & & \end{bmatrix}^{T} \begin{bmatrix} & & \\ & \Psi & \\ & & \end{bmatrix} = \begin{bmatrix} \ddots & & 0 \\ & I & \\ 0 & & \ddots \end{bmatrix} \tag{1-48}$$

From the standpoint of the vector dot product (scalar product) of covariant base vector, $\{\Psi\}_S$, and contravariant base vector, $\{\Psi'\}_r$, where the $r^{th}$ contravariant component is not a matching component to the $s^{th}$ covariant component,

$$\underline{\vec{\mathbf{b}}}_r \bullet \underline{\vec{\mathbf{b}}}_s = \begin{Bmatrix} \\ \Psi' \\ \\ \end{Bmatrix}_r^{T} \begin{Bmatrix} \\ \Psi \\ \\ \end{Bmatrix}_s = 0 \tag{1-49}$$

And when covariant and contravariant base vector components are matched up,

$$\underline{\vec{\mathbf{b}}}_r \bullet \underline{\vec{\mathbf{b}}}_r = \begin{Bmatrix} \\ \Psi' \\ \\ \end{Bmatrix}_r^{T} \begin{Bmatrix} \\ \Psi \\ \\ \end{Bmatrix}_r = 1 \tag{4-50}$$

The mode coefficients, $\Psi_{jr}$, are direction cosines of direction angles, $\alpha_{jr}$. Equation (4-50) then is the same as

$$\left\{ \Psi' \right\}_r^T \left\{ \Psi \right\}_r = \cos^2 \alpha_{1r} + \cos^2 \alpha_{2r} = 1 \qquad (4\text{-}51)$$

Revisiting the scalar product of force and displacement, beginning with

$$\{F\}^T \{X\} = \{F\}^T \{X\} \qquad (4\text{-}52)$$

Substituting into the right side of (4-52): Use the contravariant transformation of equation (4-44) for $\{F\}^T$ and the covariant transformation, equation (4-17), for $\{X\}$.

$$\left\{ \mathbf{F} \right\}^T \left\{ \mathbf{X} \right\} = \left\{ \underline{\mathbf{F'}} \right\}^T \left[ \quad \Psi' \quad \right]^T \left[ \quad \Psi \quad \right] \left\{ \underline{\mathbf{X}} \right\} \qquad (1\text{-}53)$$

Since $[\ \Psi'\ ]^T$ is the same as $[\ \Psi\ ]^{-1}$, then equation (4-53) becomes

$$\left\{ \mathbf{F} \right\}^T \left\{ \mathbf{X} \right\} = \left\{ \underline{F'} \right\}^T \left[ \quad \Psi \quad \right]^{-1} \left[ \quad \Psi \quad \right] \left\{ \underline{X} \right\} \qquad (1\text{-}54)$$

And since $[\Psi]^{-1}[\Psi]$ is the identity, $[\ \mathbf{I}\ ]$, (4-54) is just

$$\left\{\mathbf{F}\right\}^{T}\left\{\mathbf{X}\right\}=\left\{\underline{\mathbf{F}'}\right\}^{T}\left\{\underline{\mathbf{X}}\right\} \qquad (1\text{-}55)$$

We have arrived again at the now familiar statement of the scalar product of force and displacement as being represented by the row times column multiplication of the contravariant modal force and covariant modal displacement. And again we have the picture of a single contravariant force component forcing a pure modal deformation along the corresponding covariant coordinate.

## 4.8 Operators

Reflect for a moment on the operation of rotating a vector, $\vec{V}$, from one position to another and rescaling the magnitude so that the new position and magnitude is represented by the vector, $\vec{W}$. This operation is illustrated in Figure 4-22.

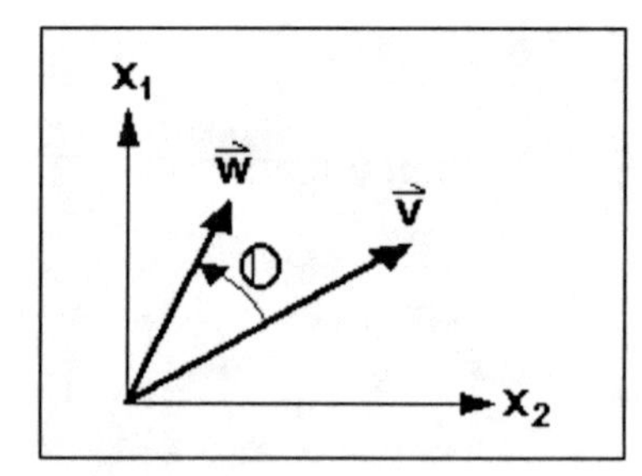


Figure 4-22. The operation of rotating the vector, V, is illustrated. The operation is indicated mathematically using the matrix operator, $\mathbb{O}$.

$$\{W\} = [O]\{V\}$$

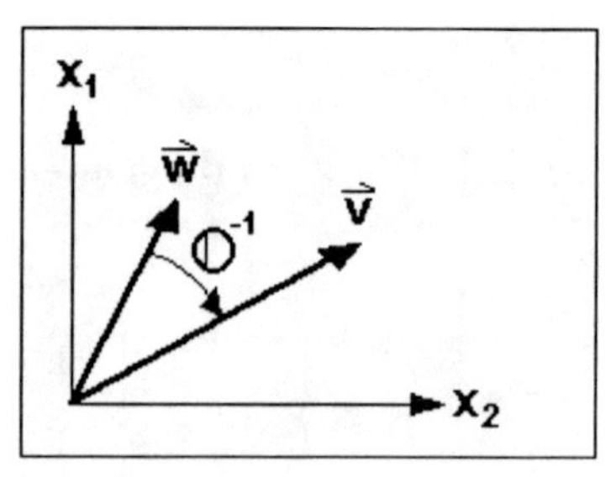


Figure 4-23. The inverse operation of rotating the vector, W, back into V is illustrated. The matrix inverse is used to represent this operation.

$$\{V\} = [O]^{-1}\{W\}$$

The operation can be represented mathematically using an "operator" matrix, $[\mathbf{O}]$, as follows.

$$\left\{ \mathbf{W} \right\} = \left[ \quad \mathbf{O} \quad \right] \left\{ \mathbf{V} \right\} \qquad (4\text{-}56)$$

This matrix equation may appear just like a coordinate transformation, but notice that the difference is that components on both sides of the equation are in the same coordinates. An operator converts components of a vector to components of a new vector within the *same* coordinate system, whereas a transformation matrix computes components of a vector for a *new* coordinate system.

The inverse operation reverses the direction (Figure 4-23), in this case carrying the vector, $\vec{\mathrm{W}}$, into the vector, $\vec{\mathrm{V}}$. The operator for this inverse operation is the matrix inverse, $[\mathbf{O}]^{-1}$.

$$\left\{ \mathbf{V} \right\} = \left[ \quad \mathbf{O} \quad \right]^{-1} \left\{ \mathbf{W} \right\} \qquad (4\text{-}57)$$

Figure 4-24 shows the same two vectors in the two degree-of-freedom modal coordinate system of Figure 4-18.

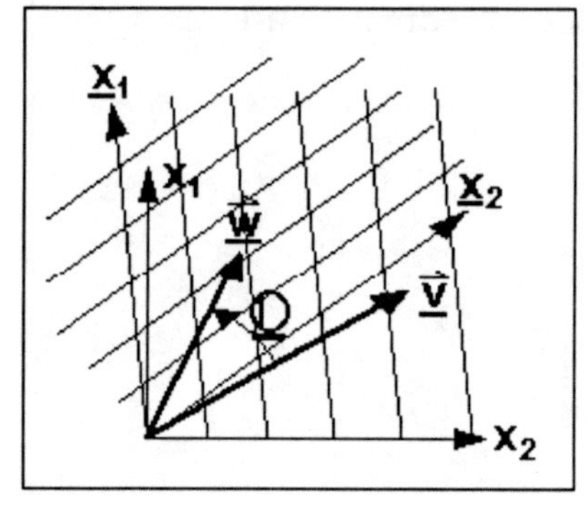


**Figure 4-24.** The operation of rotating the vector, $\underline{V}$, is illustrated. The operation is indicated mathematically using the matrix operator, $\underline{O}$ .

$$\{\underline{W}'\} = [\,\underline{O}\,]\{\underline{V}\}$$

The operation on vector, $\vec{V}$, producing vector, $\vec{W}$, can be represented in modal coordinates just as well as in physical coordinates. But now, the operator as well as the modal coordinate vector components must all be represented in modal coordinates. Clearly, the operator, $[\underline{\mathbf{O}}]$, in modal coordinates must be different than the operator, $[\mathbf{O}]$, in physical coordinates. Even though the vectors are exactly the same, the vector components being operated on are totally different for the two coordinate systems. So, the operatlion in modal coordinates for the same change from vector $\vec{V}$ to vector $\vec{W}$ is represented by the matrix equation:

$$\left\{\underline{\mathbf{W}}\right\} = \left[\quad \underline{\mathbf{O}} \quad\right]\left\{\underline{\mathbf{V}}\right\} \tag{4-58}$$

Actually, this operator is not the one that we are interested in at the moment. We really need the operation that will carry <u>*covariant*</u> modal components of one vector into <u>*contravariant*</u> modal components of another vector. This is motivated by the insight gained with equation (4-31). It was recognized there that a force vector in the direction of just one <u>*contravariant*</u> modal coordinate will produce a deformation state represented by a displacement along a

single corresponding *covariant* modal coordinate direction. So, we are looking for the operator in this case that carries the *covariant* modal components, $\{\underline{V}\}$, into the *contravariant* components of vector $\{\underline{W}'\}$. This operator can be found by adding on a transformation to equation (4-58) so that the modal coordinate components present at that stage will be carried one step further into contravariant vector components.. This type of operator will be of particular use later when an operator is needed to carry mode shape displacement vector components into contravariant modal
force vector components. Multiplying both sides of (4-58) by the metric matrix in the form of the product, $[\,\Psi\,]^T[\,\Psi\,]$,

$$\begin{bmatrix} & & \\ & \Psi & \\ & & \end{bmatrix}^T \begin{bmatrix} & & \\ & \Psi & \\ & & \end{bmatrix} \begin{Bmatrix} \\ \underline{W} \\ \\ \end{Bmatrix} =$$

$$\begin{bmatrix} & & \\ & \Psi & \\ & & \end{bmatrix}^T \begin{bmatrix} & & \\ & \Psi & \\ & & \end{bmatrix} \begin{bmatrix} & & \\ & \underline{O} & \\ & & \end{bmatrix} \begin{Bmatrix} \\ \underline{V} \\ \\ \end{Bmatrix} \qquad (4\text{-}59)$$

Since the metric matrix applied to the modal components, $\{\underline{W}\}$, transposes it into contravariant coordinates, $\{\underline{W}'\}$, the effect of the metric on the left side of the modal coordinate operator, $[\underline{\mathbf{O}}]$, is to diagonalize that operator. This is consistent with the conclusions of section 4.6 and will be proven in the next section. So, (4-59) becomes

$$\begin{Bmatrix} \\ \underline{W'} \\ \\ \end{Bmatrix} = \begin{bmatrix} \ddots & & 0 \\ & \underline{O'} & \\ 0 & & \ddots \end{bmatrix} \begin{Bmatrix} \\ \underline{V} \\ \\ \end{Bmatrix} \qquad (4\text{-}60)$$

The operation represented by (4-60) now clearly indicates carrying covariant modal components of $\{\underline{V}\}$ into contravariant modal components of $\{\underline{W}'\}$, where covariant component, $\underline{V}_1$, goes to contravariant component, $\underline{W}_1'$, then covariant component $\underline{V}_2$ goes to contravariant component, $\underline{W}_2'$, etc.

## 4.9 Transformations On Operators

Now the relationship between operators expressed in different coordinate systems will be proven. A physical coordinate operator may be transformed from physical coordinates to modal coordinates with a special transformation derived as follows. Start with the basic coordinate transformation (4-17):

$$\left\{ \mathbf{X} \right\} = \left[ \Psi \right] \left\{ \underline{\mathbf{X}} \right\} \qquad (4\text{-}61)$$

and

$$\left\{ \underline{\mathbf{X}} \right\} = \left[ \Psi \right]^{-1} \left\{ \mathbf{X} \right\} \qquad (4\text{-}62)$$

Repeating equation (4-56) which represents the operation that carries vector, $\vec{V}$, into vector, $\vec{W}$,

$$\left\{ \mathbf{W} \right\} = \left[ \quad \mathbf{O} \quad \right] \left\{ \mathbf{V} \right\} \qquad (4\text{-}63)$$

Next, replace physical coordinates on both sides of (4-63) using the transformation to modal coordinates from (4-61).

$$\begin{bmatrix} \Psi \end{bmatrix} \left\{ \underline{\mathbf{w}} \right\} = \begin{bmatrix} \mathbf{O} \end{bmatrix} \begin{bmatrix} \Psi \end{bmatrix} \left\{ \underline{\mathbf{V}} \right\} \qquad (4\text{-}64)$$

Now, pre multiply both sides of (4-64) by $[\, \Psi \,]^{-1}$

$$\begin{bmatrix} \Psi \end{bmatrix}^{-1} \begin{bmatrix} \Psi \end{bmatrix} \left\{ \underline{\mathbf{w}} \right\} =$$

$$\begin{bmatrix} \Psi \end{bmatrix}^{-1} \begin{bmatrix} \mathbf{O} \end{bmatrix} \begin{bmatrix} \Psi \end{bmatrix} \left\{ \underline{\mathbf{V}} \right\} \qquad (4\text{-}65)$$

In view of the unity product, $[\, \Psi \,]^{-1}[\, \Psi \,]$, premultiplying the l.h.s. of this equation, (4-65) reduces to

$$\left\{ \underline{\mathbf{W}} \right\} = \begin{bmatrix} \Psi \end{bmatrix}^{-1} \begin{bmatrix} \mathbf{O} \end{bmatrix} \begin{bmatrix} \Psi \end{bmatrix} \left\{ \underline{\mathbf{V}} \right\} \qquad (4\text{-}66)$$

A direct comparison of (4-66) to (4-58) shows that the transformation of the operator, $[\mathbf{O}]$, from physical coordinates to the operator, $[\underline{\mathbf{O}}]$, in *covariant* modal coordinates is

$$\left[ \quad \underline{\mathbf{O}} \quad \right] = \left[ \quad \Psi \quad \right]^{-1} \left[ \quad \mathbf{O} \quad \right] \left[ \quad \Psi \quad \right]$$

(4-67)

The operator transformation of (4-67) is known as a similarity transformation, and the two matrices, [ **O** ] and [ $\underline{\mathbf{O}}$ ], are said to be similar.

Again, we are more interested in the operator that carries the components of a vector in physical coordinates to the components expressed in *contravariant* modal coordinates. To get this transformation just premultiply both sides of (4-66) by the matrix product,
$[\Psi]^T[\Psi]$.

$$[\Psi]^T[\Psi]\{\underline{W}\} = [\Psi]^T[\Psi][\Psi]^{-1}[O][\Psi]\{\underline{V}\} \qquad (4\text{-}68)$$

The matrix product, $[\ \Psi\ ]^T[\ \Psi\ ]$, on the left side of (4-68) is recognized as the metric matrix, (4-34), which transforms the *covariant* modal coordinates, { $\underline{\mathbf{W}}$ } into *contravariant* modal coordinates, { $\underline{\mathbf{W}}$ '}. This transformation was shown in (4-38) and used again going from (4-59) to (4-60). And in view of the unity product, $[\ \Psi\ ]^{-1}[\ \Psi\ ]$, premultiplying on the r.h.s. of the equation, (4-68) becomes

$$\left\{ \underline{\mathbf{W}'} \right\} = \left[ \quad \Psi \quad \right]^{\mathbf{T}} \left[ \quad \mathbf{O} \quad \right] \left[ \quad \Psi \quad \right] \left\{ \underline{\mathbf{V}} \right\} \qquad (4\text{-}69)$$

Equation (4-60) is repeated here for direct comparison with (4-69).

$$\left\{ \underline{\mathbf{W'}} \right\} = \begin{bmatrix} \ddots & & \\ & \underline{O'} & \\ & & \ddots \end{bmatrix} \left\{ \underline{\mathbf{V}} \right\} \qquad (4\text{-}70)$$

Comparing the right hand side of (4-69) to the right hand side of (4-70), we have the equality,

$$\begin{bmatrix} \ddots & & \\ & \underline{O'} & \\ & & \ddots \end{bmatrix} = \begin{bmatrix} & & \\ & \Psi & \\ & & \end{bmatrix}^{\mathbf{T}} \begin{bmatrix} & & \\ & \mathbf{O} & \\ & & \end{bmatrix} \begin{bmatrix} & & \\ & \Psi & \\ & & \end{bmatrix} \qquad (4\text{-}71)$$

Equation (4-71) defines the transformation of an operator from physical coordinates into ***contravariant*** modal coordinates.

Notice that the operator, [ **O** ] (in physical coordinates), must be a symmetric matrix. This is seen by the following. Transpose both sides of (4-71). The left side is of course unchanged, since the off-diagonal elements are zero.

$$\begin{bmatrix} \ddots & & \\ & \underline{O'} & \\ & & \ddots \end{bmatrix} = \begin{bmatrix} & & \\ & \Psi & \\ & & \end{bmatrix}^{\mathbf{T}} \begin{bmatrix} & & \\ & \mathbf{O} & \\ & & \end{bmatrix}^{\mathbf{T}} \begin{bmatrix} & & \\ & \Psi & \\ & & \end{bmatrix} \qquad (4\text{-}72)$$

Comparing (4-72) to (4-71) shows that

$$\begin{bmatrix} & & \\ & \mathbf{O} & \\ & & \end{bmatrix} = \begin{bmatrix} & & \\ & \mathbf{O} & \\ & & \end{bmatrix}^{\mathbf{T}} \tag{4-73}$$

And this shows $[\ \mathbf{O}\ ]$ to be symmetric.

It is important to realize that we have just derived the coordinate transformation of operators for the general case, at least for any operator whose representation in modal coordinates is to signify carrying *covariant* modal components on a one-for-one basis into *contravariant* components. This transformation should apply to any operator, including the mass matrix, $[\mathrm{M}]$, stiffness matrix, $[\mathrm{K}]$ and damping matrix, $[\mathrm{C}]$. As shown above, all of these matrices must be symmetric. This is known to be the case for linear structures having proportional damping.

The mass matrix carries components of an acceleration vector into components of a force vector. Given in physical coordinates,

$$\begin{Bmatrix} \\ \mathbf{F} \\ \\ \end{Bmatrix} = \begin{bmatrix} & & \\ & \mathbf{M} & \\ & & \end{bmatrix} \begin{Bmatrix} \\ \ddot{\mathrm{X}} \\ \\ \end{Bmatrix} \tag{4-74}$$

The transformation of the mass matrix operator must conform to the prototype transformation, equation (4-71), so that

$$\begin{bmatrix} \ddots & & \\ & \underline{\mathrm{M}'} & \\ & & \ddots \end{bmatrix} = \begin{bmatrix} & & \\ & \Psi & \\ & & \end{bmatrix}^{\mathbf{T}} \begin{bmatrix} & & \\ & \mathbf{M} & \\ & & \end{bmatrix} \begin{bmatrix} & & \\ & \Psi & \\ & & \end{bmatrix} \tag{4-75}$$

Now, the sense in which mode shapes are "orthogonal" is made clear. There is a kind of orthogonality among just the regular covariant mode shapes, based on (4-75), described by the phrase, "orthogonal with respect to the mass matrix." Still, this does not mean that $[\ \Psi\ ]^T[\ \Psi\ ]$ is equal to the identity matrix, $[\ \mathbf{I}\ ]$. Nor does it mean that $[\ \Psi\ ]$ is an orthogonal transformation matrix. That would require $[\ \Psi\ ]^T$ to be equal to $[\ \Psi\ ]^{-1}$, and it is obvious that this is not the case, at least for most real structures.

Text books on matrix analysis refer to (4-75) as a congruent transformation and matrices $[\ \mathbf{M}\ ]$ and $[\ \underline{\mathbf{M}}'\ ]$ are said to be congruent. Text books usually prove that every symmetric matrix is congruent to some diagonal matrix and that a transformation matrix of the type, $[\ \Psi\ ]$, can always be found to perform the congruent transformation.

The mass operator in modal coordinates carries *covariant* modal acceleration into *contravariant* modal force as follows.

$$\left\{ \underline{\mathbf{F}}' \right\} = \begin{bmatrix} \ddots & & \\ & \underline{\mathbf{M}}' & \\ & & \ddots \end{bmatrix} \left\{ \ddot{\underline{\mathrm{X}}} \right\} \tag{4-76}$$

The results for stiffness and damping operators are the same as for the mass operator. The stiffness matrix (operator) carries a displacement vector into a force vector. Expressed in physical coordinates,

$$\left\{ \mathbf{F} \right\} = \begin{bmatrix} & & \\ & \mathbf{K} & \\ & & \end{bmatrix} \left\{ \mathbf{X} \right\} \tag{1-77}$$

The stiffness matrix in modal coordinates is obtained in the same way as the mass matrix or any other operator:

$$\begin{bmatrix} \ddots & & \\ & \underline{\mathbf{K'}} & \\ & & \ddots \end{bmatrix} = \begin{bmatrix} & & \\ & \Psi & \\ & & \end{bmatrix}^{\mathbf{T}} \begin{bmatrix} & & \\ & \mathbf{K} & \\ & & \end{bmatrix} \begin{bmatrix} & & \\ & \Psi & \\ & & \end{bmatrix} \tag{4-78}$$

And the contravariant modal force is seen as the result of the modal stiffness matrix operating on the covariant modal displacement, i.e., the modal displacement vector is rotated and scaled into a contravariant modal force vector:

$$\begin{Bmatrix} \\ \underline{\mathbf{F'}} \\ \\ \end{Bmatrix} = \begin{bmatrix} \ddots & & \\ & \underline{\mathbf{K'}} & \\ & & \ddots \end{bmatrix} \begin{Bmatrix} \\ \underline{\mathbf{X}} \\ \\ \end{Bmatrix} \tag{4-79}$$

The damping matrix operates on a velocity vector, rotating and scaling it into a force vector. The physical coordinate representation of this operation is

$$\begin{Bmatrix} \\ \mathbf{F} \\ \\ \end{Bmatrix} = \begin{bmatrix} & & \\ & \mathbf{C} & \\ & & \end{bmatrix} \begin{Bmatrix} \\ \dot{\mathrm{X}} \\ \\ \end{Bmatrix} \tag{4-80}$$

Again the operator transformation is used, this time to compute the damping operator in modal coordinates.

$$\begin{bmatrix} \ddots & & \\ & \underline{\mathbf{C'}} & \\ & & \ddots \end{bmatrix} = \begin{bmatrix} & & \\ & \Psi & \\ & & \end{bmatrix}^{\mathbf{T}} \begin{bmatrix} & & \\ & \mathbf{C} & \\ & & \end{bmatrix} \begin{bmatrix} & & \\ & \Psi & \\ & & \end{bmatrix} \tag{4-81}$$

Modal force due to modal damping is seen to result from the modal damping matrix operating on modal velocity:

$$\left\{ \underline{\mathbf{F'}} \right\} = \begin{bmatrix} \ddots & & \\ & \underline{\mathbf{C'}} & \\ & & \ddots \end{bmatrix} \left\{ \underline{\dot{\mathrm{X}}} \right\} \tag{4-82}$$

The damping matrix, [C], for many real life physical structures may not be an operator satisfying the properties of our prototype. The acid test is to see if it diagonalizes properly when applying the triple matrix product using the mode shape matrix as in (4-81). It can be shown that the [C] matrix will diagonalize only when it is a linear combination of the mass matrix, [M], and the stiffness matrix, [K]. This is naturally the case any time it is known that [C] is proportional to either [M] or [K]. A structure having a damping matrix that satisfies this condition is known as "proportionally damped." Physically, this means that the damping is distributed around the structure in the same manner as mass and/or stiffness. If the damping matrix is non proportional, it is possible to diagonalize it using a mode shape matrix that has complex mode coefficients, i.e.,

$$\Psi_{jr} = \Psi_{jr(Real)} + i\Psi_{jr(Imag)} \tag{4-83}$$

The operator transformation is then

$$\begin{bmatrix} \ddots & & \\ & \underline{\mathrm{C'}} & \\ & & \ddots \end{bmatrix} = \begin{bmatrix} & & \\ & \Psi & \\ & & \end{bmatrix}^{+} \begin{bmatrix} & & \\ & \mathbf{C} & \\ & & \end{bmatrix} \begin{bmatrix} & & \\ & \Psi & \\ & & \end{bmatrix} \tag{4-84}$$

where $[\ \Psi\ ]^{+}$ is the hermitean conjugate (transpose of the complex conjugate of $[\ \Psi\ ]$). It is not uncommon for structures in the test lab to exhibit complex mode shapes. The fact that the mode coefficient at each node point location on a structure has a real and imaginary means that there is also a mode coefficient magnitude and phase angle. This results in the observation of out of phase motion across a complex structure when viewing the vibration of a single complex mode shape at its natural frequency. A vibrating beam, for example, would display a kind of galloping motion in which the "zero node points" would appear to shift back and forth. Some engineers have referred to this phenomena as the "wandering node."

Later, when the subject of experimental modal analysis is taken up, i.e., extracting mode shapes from test data, the subject of complex modes will come up again. Some commercially available software provide the capability to obtain complex modes as part of the modal analysis process.

The modal analysis process using the Finite Element Method typically does not allow for complex mode shapes. That is because damping properties are not included in the Finite Element Model to begin with. The modal analysis is performed using only the mass and stiffness matrices. When the nonproportional damping characteristics are quite pronounced, it can be very difficult to correlate experimental and FEM models.

# CHAPTER V

# FREE VIBRATION OF A SYSTEM

Up to now we have been concerned primarily with motion and deformation patterns and their representation in different coordinate systems. The mode shapes in the previous examples were assumed. Now, the mode shapes for a freely vibrating system will be derived from the basic equations of motion. The equations of motion consist of a set of second order constant coefficient differential equations expressed in matrix form. The mass and stiffness matrices are assumed to be known.

## 5.1 Dynamical Equation Eigenvalues And Eigenvectors

The homogeneous equations of motion for the undamped case are

$$[M]\{\ddot{X}(t)\}+[K]\{X(t)\}=0 \tag{5-1}$$

Fourier transforming the equations puts them in the frequency domain.

$$[M]\{\ddot{X}(\omega)\}+[K]\{X(\omega)\}=0 \qquad (5\text{-}2)$$

There is a simple algebraic relationship between the Fourier transform of displacements, $\{X(\omega)\}$, and the Fourier transform of accelerations, $\{\ddot{X}(\omega)\}$. From equation 1-24,

$$\{\ddot{X}(\omega)\}=-\omega^2\{X(\omega)\} \qquad (5\text{-}3)$$

where $\omega$ is circular frequency (radians/sec) and $\nu$ is frequency in Hertz, so that

$$\omega=2\pi\nu \qquad (5\text{-}4)$$

Using $\lambda$ for $\omega^2$, and replacing acceleration in (5-2), we have, after factoring out the displacement and multiplying through by $[M]^{-1}$,

$$\left[-\lambda\left[{}^{\backslash}I_{\backslash}\right]+[M]^{-1}[K]\right]\{X(\omega)\}=0 \qquad (5\text{-}5)$$

Representing $[M]^{-1}[K]$ as $[B]$ and moving $\lambda[{}^{\backslash}I_{\backslash}]$ to the right hand side,

$$[B]\{X(\omega)\}=\lambda\{X(\omega)\} \qquad (5\text{-}6)$$

The set of equations represented by (5-6) have solutions in which special values of $\lambda$ match up with special sets of displacement vectors, {X}. If the [B] matrix is [n x n], and if the n equations represented by (5-6) are linearly independent, then there will be n solutions corresponding to n pairs of values for $\lambda_r$ and $\{\psi\}_r$. The rth solution requires a solution for an eigenvalue, $\lambda_r$, and an eigenvector, $\{\psi\}_r$. This is expressed by the eigenvalue equation,

$$[B]\{\psi\}_r = \lambda_r \{\psi\}_r \tag{5-7}$$

Each eigenvector corresponds physically to a mode shape of the dynamical structure as referred to in the previous sections of this text. The resonance frequency, $\nu_r$, of the rth mode is

$$\nu_r = \frac{1}{2\pi}\sqrt{\lambda_r} \tag{5-8}$$

or,

$$\nu_r = \frac{1}{2\pi}\omega_r \tag{5-9}$$

Physically, the structure will freely vibrate in any one of the oscillatory deformation patterns corresponding to one of the mode shapes, $\{\psi\}_r$, with a vibration rate or resonance frequency, $\nu_r$. A way to induce this kind of free vibration in just one specific mode shape pattern would be to enforce a static modal deformation, applying just the right static force at each physical location and direction around the structure (contravariant modal force), then suddenly release all physical forces simultaneously, allowing the structure to vibrate freely. A structure without damping would continue vibrating in the modal deformation pattern initially induced.

This method of inducing free vibration of a single mode of a structure (known as a relaxation method) has been attempted experimentally.

One difficulty for complicated structures is gaining access to sufficient degrees of freedom for attaching thin cables that are to be suddenly severed. Another problem is to cut the cables simultaneously within a sufficiently short time increment. This has been attempted using electrically fired primer cord with precise electronic timing.

Equation (5-7) can be written to explicitly include all eigenvalues and eigenvectors using the full matrix form,

$$[B][\Psi]=[\Psi]\left[{}^{\backslash}\Lambda_{\backslash}\right] \tag{5-10}$$

Here, the mode shape matrix, [ $\psi$ ], is seen to contain all of the mode shapes of the structural system, each mode shape appearing as a separate column of the matrix, [ $\psi$ ]. Keep in mind that this same matrix also plays the role of a coordinate transformation matrix from modal coordinates to physical coordinates.

## 5.2 Contravariant Eigenvectors

It is interesting to consider the concept of contravariant coordinates from the perspective of the eigenvalue/eigenvector equation. Having developed the equation for matrix [ B ],
equation (5-10), we now look at the eigenvalue/eigenvector equation for the transposed matrix, $[B]^T$. We would like to show that the eigenvalues for $[\ B\ ]^T$ are the same as for [ B ], and that the eigenvectors of $[\ B\ ]^T$ are the contravariant eigenvectors, $\{\ \Psi'\ \}$, where as before $[\ \Psi'\ ]^T\ [\ \Psi\ ] = [\ \mathbf{I}\ ]$. The physical picture here is that if you can find these eigenvectors of $[\ B\ ]^T$ and draw a picture of one of the structural deformation patterns, say $\{\ \Psi'\ \}_r$, you will see a shape that is proportional to the set of forces required to produce the covariant eigenvector deformation pattern, $\{\ \Psi\ \}_r$. This is saying that if you are a structural dynamics test engineer and would

like to know what resonance frequency and set of forces to apply to a structure to tune in a pure normal mode, just solve the eigenvalue/eigenvector equation below to find $[\,\Lambda\,]$ and $[\,\Psi'\,]$.

$$[B]^T[\Psi'] = [\Psi']\left[{}^{\backslash}\Lambda_{\backslash}\right] \tag{5-11}$$

One way of showing (5-11) is to begin by premultiplying both sides of (5-10) by the transpose matrix, $[\,\Psi'\,]^T$.

$$[\Psi']^T[\mathbf{B}][\Psi] = [\Psi']^T[\Psi]\left[{}^{\backslash}\Lambda_{\backslash}\right] \tag{5-12}$$

Note that the product on the r.h.s. of (5-12), $[\,\Psi'\,]^T[\,\Psi\,] = [\,\mathbf{I}\,]$, as developed in (5-48). So, (5-12) becomes

$$[\Psi']^T[\mathbf{B}][\Psi] = \left[{}^{\backslash}\Lambda_{\backslash}\right] \tag{5-13}$$

The diagonalization of the $[\,\mathbf{B}\,]$ matrix in equation (5-13) might at first seem in conflict with the earlier agreed upon coordinate transformation for operators, equation (5-71). Equation (5-13) makes it look like an operator has been transformed using the contravariant matrix, $[\Psi'\,]^T$, instead of $[\,\Psi\,]^T$. But, remember that the matrix, $[\,\mathbf{B}\,]$ is not a symmetric matrix, although it is the product of the two symmetric matrices,
$[\,\mathbf{M}\,]^{-1}$ and $[\,\mathbf{K}\,]$.

Now, the unity product, $[\,\Psi'\,]^T[\,\Psi\,]$, can be used to postmultiply on the r.h.s. of (5-13) without changing the equation to give

$$[\Psi']^T[B][\Psi] = \left[{}^{\backslash}\Lambda_{\backslash}\right][\Psi']^T[\Psi] \qquad (5\text{-}14)$$

Next, postmultiply both sides of (5-14) by $[\Psi]^{-1}$ :

$$[\Psi']^{\mathbf{T}}[\mathbf{B}][\Psi][\Psi]^{-1} = \left[{}^{\backslash}\Lambda_{\backslash}\right][\Psi']^{\mathbf{T}}[\Psi][\Psi]^{-1} \qquad (5\text{-}15)$$

This produces unity products, $[\Psi][\Psi]^{-1}$, as postmultiplying products on both sides of the equation, reducing (5-15) to

$$[\Psi']^{\mathbf{T}}[\mathbf{B}] = \left[{}^{\backslash}\Lambda_{\backslash}\right][\Psi']^{\mathbf{T}} \qquad (5\text{-}16)$$

Finally, the equation (5-11) result is obtained by transposing both sides of (5-16).

$$[B]^T[\Psi'] = [\Psi']\left[{}^{\backslash}\Lambda_{\backslash}\right] \qquad (5\text{-}17)$$

## 5.3 Two D.O.F. Solution

Now, the eigenvalue/eigenvector equation will be applied to a structure consisting of two masses and two springs. We wish to find two resonance frequencies and two mode shapes.

The two degree of freedom system discussed in section 4.5 (figure 4-15) will be solved using a numerical analysis procedure known as the Jacobi method. Even though this is familiar to many readers, the structural dynamicist or vibration test engineer may benefit from a review of matrix characteristics encountered here. The Jacobi method systematically reduces a matrix to diagonal form, recovering both eigenvectors (mode shapes) and eigenvalues (related to resonance frequencies) in the process.

One requirement for applying the Jacobi method is that the matrix to be analyzed must be symmetric. This presents an immediate problem for our dynamical equations. The eigenvalue problem to be solved here is equation (5-7), repeated below for convenience.

$$[B]\{\psi\}_r = \lambda_r\{\psi\}_r \tag{5-18}$$

If things were O.K., the Jacobi method would be applied to the matrix, [B]. Recall that this matrix originated from the product, $[M]^{-1}[K]$. But this matrix is generally not symmetric (even though the matrices, [M] and [K], are each symmetric) as seen below for the two degree of freedom case of figure 4-15.

$$\begin{bmatrix} \frac{1}{m_1} & 0 \\ 0 & \frac{1}{m_2} \end{bmatrix} \begin{bmatrix} k_{11} & k_{12} \\ k_{21} & k_{22} \end{bmatrix} = \begin{bmatrix} \frac{k_{11}}{m_1} & \frac{k_{12}}{m_1} \\ \frac{k_{21}}{m_2} & \frac{k_{22}}{m_2} \end{bmatrix} \tag{5-19}$$

The right side of (5-19) is symmetric only in the special case that the two masses are equal ($m_1 = m_2$). When $m_1$ does not equal m2 and the matrix is not symmetric then there is a way to get a matrix that is symmetric. First, perform a Cholesky or Crout decomposition on the mass matrix, i.e., factor symmetric [M] into the product of a lower triangular matrix and its transposed upper triangular matrix.

$$[M] = [L][L]^T \tag{5-20}$$

The Cholesky decomposition (or Crout) is indicated here in order to represent the procedure that could be followed for the general case where large nondiagonal mass matrices could be encountered. Factoring our simple two by two matrix with zeros in the off diagonals is done by inspection:

$$\begin{bmatrix} m_1 & 0 \\ 0 & m_2 \end{bmatrix} = \begin{bmatrix} \sqrt{m_1} & 0 \\ 0 & \sqrt{m_2} \end{bmatrix} \begin{bmatrix} \sqrt{m_1} & 0 \\ 0 & \sqrt{m_2} \end{bmatrix} \tag{5-21}$$

And the inverse, $[M]^{-1}$, is

$$[M]^{-1} = \begin{bmatrix} \frac{1}{m_1} & 0 \\ 0 & \frac{1}{m_2} \end{bmatrix} = \begin{bmatrix} \frac{1}{\sqrt{m_1}} & 0 \\ 0 & \frac{1}{\sqrt{m_2}} \end{bmatrix} \begin{bmatrix} \frac{1}{\sqrt{m_1}} & 0 \\ 0 & \frac{1}{\sqrt{m_2}} \end{bmatrix} \tag{5-22}$$

Or,

$$[M]^{-1} = [L]^{-T}[L]^{-1} \tag{5-23}$$

Now, start again with equation (5-5). Substitute the right side of (5-23) in place of $[M]^{-1}$.

$$\left[-\lambda + [\mathbf{L}]^{-\mathbf{T}}[\mathbf{L}]^{-1}[\mathbf{K}]\right]\{\mathbf{X}\} = 0 \qquad (5\text{-}24)$$

Pre multiply by $[\mathrm{L}]^{\mathrm{T}}$ and also sandwich in the identity matrix in the form of $[\mathrm{I}] = [\mathrm{L}]^{-\mathrm{T}}[\mathrm{L}]^{\mathrm{T}}$.

$$[\mathbf{L}]^{\mathbf{T}}\left[-\lambda + [\mathbf{L}]^{-\mathbf{T}}[\mathbf{L}]^{-1}[\mathbf{K}]\right][\mathbf{L}]^{-\mathbf{T}}[\mathbf{L}]^{\mathbf{T}}\{\mathbf{X}\} = 0 \qquad (5\text{-}25)$$

Multiplying through,

$$[-\lambda\,[\mathbf{I}] + [\mathbf{L}]^{-1}[\mathbf{K}][\mathbf{L}]^{-\mathrm{T}}][\mathbf{L}]^{\mathrm{T}}\{\mathbf{X}\} = 0 \qquad (5\text{-}26)$$

Notice now that (5-26) implies a coordinate transformation that would produce a standard form of the eigenvalue/eigenvector equation. To put (5-26) in a slightly more recognizable form, a change of variables will be used. Let's first assign a new matrix, $[\boldsymbol{\Phi}]$, as follows: $[\boldsymbol{\Phi}] = [\mathbf{L}]^{-\mathrm{T}}$, or $[\mathbf{L}] = [\boldsymbol{\Phi}]^{-\mathrm{T}}$. Equation (5-26) is rewritten as

$$[-\lambda\,[\mathbf{I}] + [\boldsymbol{\Phi}]^{\mathrm{T}}[\mathbf{K}][\boldsymbol{\Phi}]][\boldsymbol{\Phi}]^{-1}\{\mathbf{X}\} = 0 \qquad (5\text{-}27)$$

Equation (5-27) in this form makes it easy to spot the coordinate transformation on $\{\mathbf{X}\}$ and the coordinate transformation of the operator, $[\mathbf{K}]$. A new set of coordinates, $\{\underline{\underline{\mathbf{X}}}\}$ (double underline), will be used as the coordinates in which the (5-27) equations are expressed:

$$[-\lambda[\mathbf{I}] + [\underline{\underline{\mathbf{K}}}]]\{\underline{\underline{\mathbf{X}}}\} = 0 \qquad (5\text{-}28)$$

or,

$$[\underline{\underline{\mathbf{K}}}]\{\underline{\underline{X}}\} = \lambda\{\underline{\underline{X}}\} \qquad (5\text{-}29)$$

where the coordinate transformations that have been used are

$$\{ X \} = [ \Phi ]\{ \underline{\underline{X}} \} \qquad (5\text{-}30)$$

and

$$[\underline{\underline{\mathbf{K}}}] = [\Phi]^T[\mathbf{K}][\Phi] \qquad (5\text{-}31)$$

So, now we have a new eigenvalue problem for the ***symmetric*** matrix, $[\underline{\underline{\mathbf{K}}}]$. Once again the allowable solutions to the equations, (5-29), for $\{\underline{\underline{X}}\}$ and $\lambda$ are the eigenvectors and eigenvalues as implied in the eigenvalue/eigenvector equation for the rth eigenvector and eigenvalue. But this time the eigenvectors will be defined in the interim coordinate system, $\underline{\underline{X}}$. In other words the mode coefficients for a given mode shape will be numbers associated with the coordinates, $\underline{\underline{X}}_1$ and $\underline{\underline{X}}_2$. Accordingly, the mode shapes will be expressed in the new coordinate system as $[ \underline{\underline{\psi}} ]$. The eigenvalue/eigenvector equation for the new coordinates will be

$$[\underline{\underline{\mathbf{K}}}]\{ \underline{\underline{\psi}} \}_r = \lambda_r\{ \underline{\underline{\psi}} \}_r \qquad (5\text{-}32)$$

The eigenvalues are scalar invariants and have not been altered by the change in coordinates. Again, we will not directly get the physical coordinate mode coefficients, $\{\Psi\}_r$, that we're ultimately looking for. With (5-32) we of course have different eigenvector component values for $\{\underline{\underline{\psi}}\}_r$ than we would have had for $\{\psi\}_r$. This must be the case since $\{\underline{\underline{\psi}}\}_r$ and $\{\psi\}_r$ both represent the same physical eigenvector, but $\underline{\underline{X}}$ is a different coordinate system than the coordinates, X, used for $\{\psi\}_r$ in equation (5-7). Thus, the operator in the new coordinate system is carrying out the same operation on the same vector, just in a different coordinate system. That's important because we want the eigenvectors, $\{\underline{\underline{\psi}}\}_r$, of (5-32) to correspond to the real world mode shapes that vibrate at the resonance frequencies given by the eigenvalues, even though, after getting the values for $\{\underline{\underline{\psi}}\}_r$,we must do a coordinate transformation to see what they look like in the real world X coordinates, i.e., $\{\psi\}_1$ and $\{\psi\}_2$. Actually, for the special 2DOF example used here, the new set of coordinates, $\underline{\underline{X}}$, point in the same direction as the X coordinates. The $\{\underline{\underline{X}}\}$ vectors just have different lengths than the {X} vectors. This will not be the case in general. It can be seen to be the case in this 2DOF example by remembering that the columns of the transformation matrix, $[\Phi]$, are the base vectors of the new coordinate system, $\underline{\underline{X}}$. And since the eigenvector base vector columns of matrix, $[\Phi]$, came originally from $[\mathbf{L}]^{-T}$, we can substitute $[\mathbf{L}]^{-T}$ for $[\Phi]$ in equation (5-30) to obtain

$$\{X\} = [L]^{-T}\{\underline{\underline{X}}\} \tag{5-33}$$

Thus, reading coefficients directly from the matrices of equation (5-22), the base vectors are then

$$\underline{\underline{\vec{\mathbf{b}}_1}} \Leftrightarrow \begin{Bmatrix} \mathbf{x}_1 \\ \mathbf{x}_2 \end{Bmatrix} = [\mathbf{L}]^{-T} \begin{Bmatrix} 1 \\ 0 \end{Bmatrix} = \begin{Bmatrix} \dfrac{1}{\sqrt{m_1}} \\ 0 \end{Bmatrix} \tag{5-34}$$

and

$$\underline{\underline{\vec{\mathbf{b}}_2}} \Leftrightarrow \begin{Bmatrix} \mathbf{x}_1 \\ \mathbf{x}_2 \end{Bmatrix} = [\mathbf{L}]^{-T} \begin{Bmatrix} 0 \\ 1 \end{Bmatrix} = \begin{Bmatrix} 0 \\ \dfrac{1}{\sqrt{m_2}} \end{Bmatrix} \tag{5-35}$$

Equations (5-34) and (5-35) indicate that the coordinate transformation is just rescaling the original X coordinates without changing the direction of the coordinates. This is illustrated in Figure 5-1.

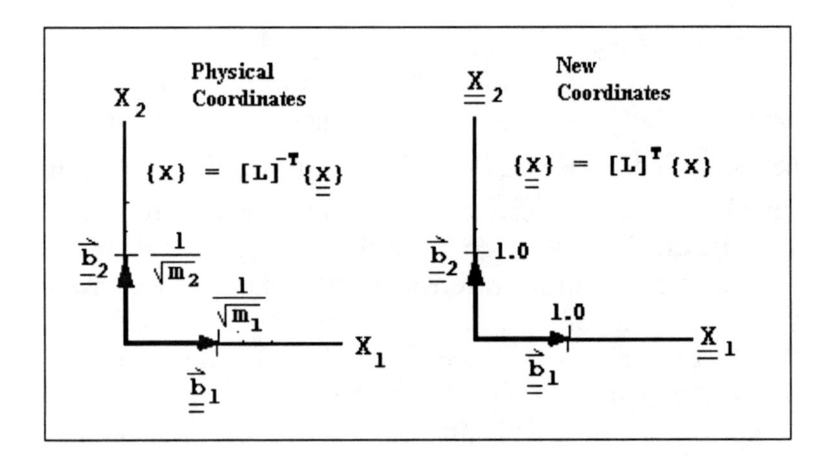


**Figure 5-1. New interim coordinates, $\underline{\underline{X}}$, compared to original physical coordinates, X.**

The matrix, $[\underline{\underline{\mathbf{K}}}]$, is now symmetric, allowing application of the Jacobi method. Once the eigenvectors, $[\,\underline{\underline{\psi}}\,]$, of $[\underline{\underline{\mathbf{K}}}]$ have been found, equation (5-30) can be used to transform the eigenvectors from the $\underline{\underline{X}}$ coordinate system into components of the X coordinate system. In terms of the eigenvector components for the two coordinate systems, (5-30) means that

$$\{\,\psi\,\}_r = [\Phi]\{\,\underline{\underline{\psi}}\,\}_r \tag{5-36}$$

Now, back to the Jacobi method. The Jacobi strategy for a large symmetric matrix is to multiply that matrix by the product of a number of orthogonal transformation matrices, where each individual transformation sets two diagonally opposite off diagonal elements to zero. Such a transformation matrix, $[T]_{ij}$ is easily found. It would consist of an identity matrix modified to include systematically selected diagonal and off-diagonal elements as indicated below.

$$[T]_{ij} = \begin{bmatrix} 1 & 0 & \cdots & & & & \cdots & 0 \\ 0 & \ddots & & & & & \iddots & \vdots \\ \vdots & & & & & & & \\ & & c_{ii} & \cdots & s_{ij} & & & \\ & & & \ddots & & & & \\ & & \vdots & 1 & \vdots & & & \\ & & & & \ddots & & & \\ & & -s_{ji} & \cdots & c_{jj} & & & \\ & & & & & \ddots & & \\ \vdots & \iddots & & & & & 1 & \vdots \\ 0 & & \cdots & & & \cdots & & 1 \end{bmatrix} \tag{5-37}$$

The selected elements have the effect of a two-by-two transformation on the affected elements of the matrix to be transformed. The two-by-two matrix is:

$$\begin{bmatrix} c_{ii} & s_{ij} \\ -s_{ji} & s_{jj} \end{bmatrix} = \begin{bmatrix} \cos(\theta_{ii}) & \sin(\theta_{ij}) \\ -\sin(\theta_{ji}) & \cos(\theta_{jj}) \end{bmatrix} \tag{5-38}$$

where the indicated sub matrix constitutes an orthogonal rotational matrix with elements that are cosines and sines. The element, c, represents cosine($\theta$) and the element, s, represents sine($\theta$). Of course the right values of $\theta_{ij}$ must be found for each new $_{ij}$th sub matrix. But once they have been found the result is the annihilation of all off-diagonal elements. Only the diagonal retaines non-zero values and these are the eigenvalues.

Since the matrix must be transformed as an operator, the actual transformation to be used is of the form derived in equation (5-71). That means a string of product matrices to the left and also to the right of the matrix to be transformed is required. So, the matrix, $[\underline{\underline{\mathbf{K}}}]$, is diagonalized with the sequence of two-by-two orthogonal transformations,

$$\left[\ddots\underline{\mathbf{K}}\ddots\right] = \sum_{i} \sum_{j} [\mathbf{T}]^{\mathrm{T}}_{ji} [\underline{\underline{\mathbf{K}}}] [\mathbf{T}]_{ij} \tag{5-39}$$

The sequence of transformations has the same effect as would the transformation implied by the eigenvalue equation of (5-32), i.e.,

$$[\ \backslash \underline{\mathbf{K}} \backslash\ ] = [\ \underline{\underline{\psi}}\ ]^{\mathbf{T}} [\ \underline{\underline{\mathbf{K}}}\ ][\ \underline{\underline{\psi}}\ ] \tag{5-40}$$

Notice that the transformation is from the interim $\underline{\underline{X}}$ coordinates to the modal coordinates, $\underline{X}$ , and

$$\{\ \mathbf{X}\ \} = [\ \underline{\underline{\psi}}\ ][\ \underline{\underline{X}}\ ] \tag{5-41}$$

so that once the sequence of transformations of (5-39) is found, it automatically yields the eigenvectors, $\{\ \underline{\underline{\psi}}\ \}$, in the interim coordinates, $\underline{\underline{X}}$ .

Actually, we would not be using the Jacobi method if our matrix was large. Jacobi is not as efficient for large matrices as other methods. One might first reduce the matrix to tridiagonal form using a method such as Givens , modified Givens or Householder. In our present example with the simple two by two matrix, [K], the matrix is already in the simplest tridiagonal form possible. So, the transformation matrix is just the sub matrix of cosines and sines.

For the simple two degree of system it remains only to find the angle, $\theta$, such that the matrix, $[\underline{\underline{\mathbf{K}}}]$ is diagonalized when transformed using the simple coordinate transformation on the operator, $[\underline{\underline{\mathbf{K}}}]$, as given in equation (5-40). Again, just one transformation is needed for our special two DOF case instead of a sequence of transformations indicated in equation (5-39). And once the transformation matrix is found, it turns out to be the eigenvector matrix, [ $\underline{\psi}$ ], corresponding to the eigenvalue equation, (5-32).

Writing out the equation (5-40) matrices in detail, using $\underline{\underline{\mathbf{K}}}_{12} = \underline{\underline{\mathbf{K}}}_{21}$,

$$\begin{bmatrix} \underline{\mathbf{K}}_{11} & 0 \\ 0 & \underline{\mathbf{K}}_{22} \end{bmatrix} = \begin{bmatrix} c & -s \\ s & c \end{bmatrix} \begin{bmatrix} \underline{\underline{\mathbf{K}}}_{11} & \underline{\underline{\mathbf{K}}}_{12} \\ \underline{\underline{\mathbf{K}}}_{21} & \underline{\underline{\mathbf{K}}}_{22} \end{bmatrix} \begin{bmatrix} c & s \\ -s & c \end{bmatrix} \tag{5-42}$$

Carrying out the multiplication for (5-42),

$$\begin{bmatrix} \underline{\mathbf{K}}_{11} & 0 \\ 0 & \underline{\mathbf{K}}_{22} \end{bmatrix} =$$

$$\begin{bmatrix} (c\underline{\underline{\mathbf{K}}}_{11}c - c\underline{\underline{\mathbf{K}}}_{12}s - s\underline{\underline{\mathbf{K}}}_{12}c + s\underline{\underline{\mathbf{K}}}_{22}s) & (c\underline{\underline{\mathbf{K}}}_{11}s + c\underline{\underline{\mathbf{K}}}_{12}c - s\underline{\underline{\mathbf{K}}}_{12}s - s\underline{\underline{\mathbf{K}}}_{22}c) \\ (s\underline{\underline{\mathbf{K}}}_{11}c - s\underline{\underline{\mathbf{K}}}_{12}s + c\underline{\underline{\mathbf{K}}}_{12}c - c\underline{\underline{\mathbf{K}}}_{22}s) & (s\underline{\underline{\mathbf{K}}}_{11}s + s\underline{\underline{\mathbf{K}}}_{12}c + c\underline{\underline{\mathbf{K}}}_{12}s + c\underline{\underline{\mathbf{K}}}_{22}c) \end{bmatrix} \tag{5-43}$$

The objective at this point is to get an equation that allows a solution for the angle, θ. So, it is only necessary to pick an element, say the upper right corner element of the right side of (5-43) and set it equal to the corresponding element on the left side, $\underline{\underline{\mathbf{K}}}_{12}$, which is an off diagonal zero.

$$c\underline{\underline{\mathbf{K}}}_{11}s + c\underline{\underline{\mathbf{K}}}_{12}c - s\underline{\underline{\mathbf{K}}}_{12}s - s\underline{\underline{\mathbf{K}}}_{22}c = 0 \tag{5-44}$$

Factor (5-44) as follows.

$$(cs)(\underline{\underline{\mathbf{K}}}_{11} - \underline{\underline{\mathbf{K}}}_{22}) + (c^2 - s^2)\underline{\underline{\mathbf{K}}}_{12} = 0 \qquad (5\text{-}44)$$

Dividing through by factored terms and moving one expression to the other side of (5-44) results in

$$\frac{\mathbf{c}^2 - \mathbf{s}^2}{\mathbf{cs}} = \frac{\underline{\underline{\mathbf{K}}}_{22} - \underline{\underline{\mathbf{K}}}_{11}}{\underline{\underline{\mathbf{K}}}_{12}} \qquad (5\text{-}45)$$

Introduce a common factor of two in the denominator to obtain a standard identity form.

$$\frac{\cos^2\theta - \sin^2\theta}{2\cos\theta\sin\theta} = \frac{\underline{\underline{\mathbf{K}}}_{22} - \underline{\underline{\mathbf{K}}}_{11}}{2\underline{\underline{\mathbf{K}}}_{12}} \qquad (5\text{-}46)$$

Substituting standard trig identities into the numerator and denominator of (5-46)

$$\frac{\cos(2\theta)}{\sin(2\theta)} = \frac{\underline{\underline{\mathbf{K}}}_{22} - \underline{\underline{\mathbf{K}}}_{11}}{2\underline{\underline{\mathbf{K}}}_{12}} \qquad (5\text{-}47)$$

Recognizing the cotangent on the left side of (5-47), the angle, θ, can be solved as

$$\theta = \frac{1}{2}\tan^{-1}\left(\frac{2\underline{\underline{\mathbf{K}}}_{12}}{\underline{\underline{\mathbf{K}}}_{22} - \underline{\underline{\mathbf{K}}}_{11}}\right) \qquad (5\text{-}48)$$

To compute a value for angle, θ, elements of the $[\underline{\underline{\mathbf{K}}}]$ matrix are needed. These are obtained from carrying out the indicated matrix multiplication of equation (5-28).

$$\begin{bmatrix} \underline{\underline{\mathbf{K}}}_{11} & \underline{\underline{\mathbf{K}}}_{12} \\ \underline{\underline{\mathbf{K}}}_{21} & \mathbf{K}_{22} \end{bmatrix} = \begin{bmatrix} \dfrac{1}{\sqrt{m_1}} & 0 \\ 0 & \dfrac{1}{\sqrt{m_2}} \end{bmatrix} \begin{bmatrix} k_{11} & k_{12} \\ k_{21} & k_{22} \end{bmatrix} \begin{bmatrix} \dfrac{1}{\sqrt{m_1}} & 0 \\ 0 & \dfrac{1}{\sqrt{m_2}} \end{bmatrix} \tag{5-49}$$

$$\begin{bmatrix} \underline{\underline{\mathbf{K}}}_{11} & \underline{\underline{\mathbf{K}}}_{12} \\ \underline{\underline{\mathbf{K}}}_{21} & \underline{\underline{\mathbf{K}}}_{22} \end{bmatrix} = \begin{bmatrix} \dfrac{k_{11}}{m_1} & \dfrac{k_{12}}{\sqrt{m_1 m_2}} \\ \dfrac{k_{12}}{\sqrt{m_1 m_2}} & \dfrac{k_{22}}{m_2} \end{bmatrix} \tag{5-50}$$

So, the angle, θ, is

$$\theta = \frac{1}{2}\tan^{-1}\left[\frac{2\left(\dfrac{\mathrm{k}_{12}}{\sqrt{\mathrm{m}_1\mathrm{m}_2}}\right)}{\dfrac{\mathrm{k}_{22}}{\mathrm{m}_2} - \dfrac{\mathrm{k}_{11}}{\mathrm{m}_1}}\right] \tag{5-51}$$

The following mass and stiffness values will be given for this example:

$$m_1 = 0.1 \text{ LB}_m \tag{5-52}$$

$$k_1 = 100 \text{ LB/in} \tag{5-53}$$

$$m_2 = 0.01 \text{ LB}_m \tag{5-54}$$

$$k_2 = 1000 \text{ LB/in} \tag{5-55}$$

The stiffness matrix for equation (5-1) is

$$\begin{bmatrix} k_{11} & k_{12} \\ k_{21} & k_{22} \end{bmatrix} = \begin{bmatrix} (k_1 + k_2) & -k_2 \\ -k_2 & k_2 \end{bmatrix} \tag{5-56}$$

Putting the numbers from (5-53) and (5-55) into (5-56),

$$\begin{bmatrix} k_{11} & k_{12} \\ k_{21} & k_{22} \end{bmatrix} = \begin{bmatrix} 1100 & -1000 \\ -1000 & 1000 \end{bmatrix} \tag{5-57}$$

Substituting these numbers into equation (5-51) gives the value for the angle, θ.

$$\theta = \frac{1}{2}\tan^{-1}\left[\frac{2\left(\dfrac{-1000}{\sqrt{(0.1)(0.01)}}\right)}{\left(\dfrac{1000}{.01}\right) - \left(\dfrac{1100}{.1}\right)}\right] \tag{5-58}$$

$$\theta = 17.69926 \tag{5-59}$$

Computing values for cosine(θ) and sine(θ),

$$\mathrm{Cos}(\theta) = .952665 \tag{5-60}$$

$$\mathrm{Sin}(\theta) = .304021 \tag{5-61}$$

The eigenvector matrix, [ $\underline{\underline{\Psi}}$ ], in $\underline{\underline{X}}$ coordinates is

$$[\underline{\underline{\Psi}}] = \begin{bmatrix} \cos\theta & \sin\theta \\ -\sin\theta & \cos\theta \end{bmatrix} \tag{5-62}$$

$$[\underline{\underline{\Psi}}] = \begin{bmatrix} .952665 & .304021 \\ -.304021 & .952665 \end{bmatrix} \tag{5-63}$$

The eigenvectors in $\underline{\underline{X}}$ coordinates are the columns of the (5-63) matrix. But we really want to see the mode shapes in the original physical X coordinates. The mode shape matrix, [ Ψ ], in physical coordinates, X, are computed using the transformation from equation (5-36).

$$\{ \psi \}_r = [\Phi]\{ \underline{\underline{\psi}} \}_r \tag{5-64}$$

Or, using $[\,\mathbf{L}\,]^{-T}$, the matrix from which [ Φ ] was defined earlier (going from equation 5-27 to 5-28),

$$\{\Psi\}_r = [\mathbf{L}]^{-\mathbf{T}}\{\underline{\underline{\Psi}}\}_r \tag{5-65}$$

Expressing (5-65) in full matrix form,

$$\left[\Psi\right]=\left[L\right]^{-T}\left[\underline{\underline{\Psi}}\right] \tag{5-66}$$

Using (5-66) the mode shape coefficients in physical coordinates are computed as follows.

$$\left[\Psi\right]=\begin{bmatrix} \dfrac{1}{\sqrt{m_1}} & 0 \\ 0 & \dfrac{1}{\sqrt{m_2}} \end{bmatrix}\begin{bmatrix} \underline{\underline{\Psi}}_{11} & \underline{\underline{\Psi}}_{12} \\ \underline{\underline{\Psi}}_{21} & \underline{\underline{\Psi}}_{22} \end{bmatrix} \tag{5-67}$$

Using the numbers from equations (5-52), (5-54) and (5-63),

$$\left[\Psi\right]=\begin{bmatrix} 3.162278 & 0 \\ 0 & 10 \end{bmatrix}\begin{bmatrix} .952665 & .304021 \\ -.304021 & .952665 \end{bmatrix} \tag{5-68}$$

$$\left[\Psi\right]=\begin{bmatrix} 3.01259 & .961399 \\ -3.04021 & 9.52665 \end{bmatrix} \tag{5-69}$$

Eigenvalue computations are seen by writing the full matrix form of the eigenvalue equation, (5-32).

$$\left[\underline{\underline{\mathbf{K}}}\right]\left[\underline{\underline{\Psi}}\right]=\left[\underline{\underline{\Psi}}\right]\left[{}^{\ddots}\Lambda_{\ddots}\right] \tag{5-70}$$

Premultiplying both sides by $[\underline{\underline{\Psi}}]^T$, noting that $[\underline{\underline{\Psi}}]$ is orthogonal so that $[\underline{\underline{\Psi}}]^T = [\underline{\underline{\Psi}}]^{-1}$,

$$[\underline{\underline{\Psi}}]^T [\underline{\underline{\mathbf{K}}}][\underline{\underline{\Psi}}] = [\,{}^{\ddots}\Lambda_{\ddots}\,] \tag{5-71}$$

But we already know from (5-40) that the left side of (5-71) diagonalizes $[\underline{\underline{\mathbf{K}}}]$, so (5-71) becomes

$$[\,{}^{\ddots}\underline{\underline{\mathbf{K}}}_{\ddots}\,] = [\,{}^{\ddots}\Lambda_{\ddots}\,] \tag{5-72}$$

Putting numbers into the $[\underline{\underline{\mathbf{K}}}]$ matrix, guided by (5-50),

$$\begin{bmatrix} \underline{\underline{\mathbf{K}}}_{11} & \underline{\underline{\mathbf{K}}}_{12} \\ \underline{\underline{\mathbf{K}}}_{21} & \underline{\underline{\mathbf{K}}}_{22} \end{bmatrix} = \begin{bmatrix} \dfrac{k_{11}}{m_1} & \dfrac{k_{12}}{\sqrt{m_1 m_2}} \\ \dfrac{k_{12}}{\sqrt{m_1 m_2}} & \dfrac{k_{22}}{m_2} \end{bmatrix} = \begin{bmatrix} \dfrac{100}{.1} & \dfrac{-1000}{\sqrt{(.1)(.01)}} \\ \dfrac{-1000}{\sqrt{(.1)(.01)}} & \dfrac{1000}{.01} \end{bmatrix}$$

(5-73)

$$\begin{bmatrix} \underline{\underline{\mathbf{K}}}_{11} & \underline{\underline{\mathbf{K}}}_{12} \\ \underline{\underline{\mathbf{K}}}_{21} & \underline{\underline{\mathbf{K}}}_{22} \end{bmatrix} = \begin{bmatrix} 1000 & -3.16228(10^4) \\ -3.16228(10^4) & 10^5 \end{bmatrix} \tag{5-74}$$

Computing the eigenvalues, $[\,{}^{\ddots}\Lambda_{\ddots}\,]$, from (5-71) using numbers from (5-74) and (5-63),

$$\left[\ \Lambda\ \right]=\begin{bmatrix} .952665 & -.304021 \\ -.304021 & .952665 \end{bmatrix}\begin{bmatrix} 1000 & -3.16228(10^4) \\ -3.16228(10^4) & 10^5 \end{bmatrix}\begin{bmatrix} .952665 & -.304021 \\ -.304021 & .952665 \end{bmatrix}$$

(5-75)

$$\begin{bmatrix} \lambda_1 & 0 \\ 0 & \lambda_2 \end{bmatrix}=\begin{bmatrix} 908.34 & 0 \\ 0 & 1.1009(10^5) \end{bmatrix} \quad (5\text{-}76)$$

The frequency for mode number one is

$$\nu_1=\frac{1}{2\pi}\sqrt{\lambda_1}=\frac{1}{2\pi}\sqrt{908.34}=4.797 \text{ Hz} \quad (5\text{-}77)$$

And for mode number two,

$$\nu_2=\frac{1}{2\pi}\sqrt{\lambda_2}=\frac{1}{2\pi}\sqrt{1.1009(10^5)}=52.81 \text{ Hz} \quad (5\text{-}78)$$

The resulting mode shapes, (5-69), and resonance frequencies, (5-77) and (5-78), are those describing the two D.O.F. system of Figures 4-15, 4-16 and 4-17 and the modal coordinate system of Figure 4-18. It is clear from the mode shape coefficients in the matrix of (5-68) and the Figure 4-18 modal coordinates (whose directions are specified by the mode coefficients) that the modal coordinates for our two D.O.F. system are certainly not perpendicular.

## 5.4 Two D.O.F. Followup

Some follow-up computations on the two D.O.F. system should be covered. First, to double check orthogonality of the mode shape matrix "with respect to the mass matrix:"

$$\left[\ddots \underline{M} \ddots\right]=[\Psi]^T[M][\Psi] \tag{5-79}$$

$$\begin{bmatrix} \underline{m}_1 & 0 \\ 0 & \underline{m}_2 \end{bmatrix}=\begin{bmatrix} 3.01259 & -3.04021 \\ .961399 & 9.52665 \end{bmatrix}\begin{bmatrix} .1 & \\ & .01 \end{bmatrix}\begin{bmatrix} 3.01259 & .961399 \\ -3.04021 & 9.52665 \end{bmatrix} \tag{5-80}$$

$$\begin{bmatrix} \underline{m}_1 & 0 \\ 0 & \underline{m}_2 \end{bmatrix}\cong\begin{bmatrix} 1.0 & 0 \\ 0 & 1.0 \end{bmatrix} \tag{5-81}$$

Next, check to see that diagonalizing the stiffness matrix produces eigenvalues. Start again with the original differential equations of motion in the frequency domain.

$$[-\lambda[M]+[K]]\{X\}=0 \tag{5-82}$$

Pre multiplying by $[\Psi]^T$ and sandwiching in the identity matrix in the form of $[I]=[\Psi][\Psi]^{-1}$,

$$[\Psi]^T[-\lambda[M]+[K]][\Psi][\Psi]^{-1}\{X\}=0 \tag{5-83}$$

$$\left[-\lambda[\Psi]^T[M][\Psi]+[\Psi]^T[K][\Psi]\right]\{\underline{X}\}=0 \tag{5-84}$$

Since the mass matrix diagonalizes to the identity, [ **I** ], this results in

$$[\Psi]^{\mathrm{T}}[\mathrm{K}][\Psi]=\left[\ddots\Lambda\ddots\right] \tag{5-85}$$

Or,

$$\left[\ddots\underline{\mathrm{K}}\ddots\right]=\left[\ddots\Lambda\ddots\right] \tag{5-86}$$

Equation (5-86) emphasizes that when mode shapes are normalized so that modal mass is one, [ M ] = **[ I ]**, then the modal stiffness values are equal to the corresponding eigenvalues. Putting the numbers into (5-85) verifies this is the case for the present two D.O.F. example ( [ Ψ ] from 5-69 and [K] from 5-57).

$$\left[\ddots\Lambda\ddots\right]=\begin{bmatrix}3.01259 & -3.04021\\ .961399 & 9.52665\end{bmatrix}\begin{bmatrix}1100 & -1000\\ -1000 & 1000\end{bmatrix}\begin{bmatrix}3.01259 & .961399\\ -3.04021 & 9.52665\end{bmatrix} \tag{5-87}$$

$$\left[\ddots\Lambda\ddots\right]=\begin{bmatrix}908.34 & 0\\ 0 & 1.1009(10^5)\end{bmatrix} \tag{5-88}$$

These eigenvalues agree with those computed in (5-76).

Finally, a computation of the metric. The metric, [G], defined in equation (4-34), is computed as (notice $G = G^T$):

$$[G]=[\Psi]^T[\Psi] \tag{5-89}$$

$$[G]=\begin{bmatrix} 3.01259 & -3.04021 \\ .961399 & 9.52665 \end{bmatrix}\begin{bmatrix} 3.01259 & .961399 \\ -3.04021 & 9.52665 \end{bmatrix} \tag{5-90}$$

$$[G]=\begin{bmatrix} 18.31858 & -26.0667 \\ -26.0667 & 91.68135 \end{bmatrix} \tag{5-91}$$

## 5.5 Operator Interpretation of The Eigenvalue Equation

A final thought on the interpretation of the eigenvalue equation, (5-7), in the context of operators: Consider the matrix, [B], as the product of the two operators that originally defined [B] going from (5-5) to (5-6).

$$[B]=[M]^{-1}[K] \tag{5-92}$$

So, now the eigenvalue equation, (5-7), is interpreted in terms of the physical operations implied by the two operators, $[M]^{-1}$ and [K].

$$[M]^{-1}[K]\{\Psi\}_r=\lambda_r\{\Psi\}_r \tag{5-93}$$

And we can picture a short sequence of operations. The first picture is the matrix operator, [K], operating on a specific displacement state vector, $\{ \Psi \}_r$, rotating and scaling that vector into a contravariant force vector, $\{F\}_{r'}$. This brings the eigenvalue equation to the condition,

$$[M]^{-1}\{F\}_{r'} = \lambda_r \{\Psi\}_r \tag{5-94}$$

Now, (5-94) is saying that the operator, $[M]^{-1}$, must perform an inverse operation that rotates the *contravariant force vector*, $\{F\}_{r'}$, right back into the opposite direction, converting it into an *acceleration vector*, which points in the direction of the original displacement vector. If that acceleration vector is rescaled by dividing by $\omega_r^2$ , i.e., $\lambda_r$, then the resulting vector is the same as the original displacement vector, $\{ \Psi \}_r$.

To get the full flavor of this operator interpretation, it must be realized that these two operators, $[M]^{-1}$ and [K] will not work this way for just any starting displacement vector and just any frequency. In a state of free vibration, there are no external forces acting on the mass. The spring force is applied to the mass, and it is reacted by the inertia of the mass. This means that the only frequencies, displacement state vectors and acceleration vectors possible are those pairs that allow the sequence of operations described above.

Under forced vibrations the operator relationship, hence the eigenvalue equation, is still relevant. A pure resonance condition occurs in a real system having proportional damping only when the same operator sequence balances out. That is, when under a particular set of externally applied loads (corresponding to a unique contravariant modal force) and oscillatory frequency, the stiffness operator finds just the right displacement vector,

$\{\Psi\}_r$ , for conversion into just the right contravariant force vector, $\{F\}_{r'}$ , such that the available inverse mass operator brings the force vector back to an acceleration vector which points in the same direction as the original displacement state vector, $\{ \Psi \}_r$ .

This happens because at resonance the reaction force of the modal spring, $\underline{K}_r$, is equal and opposite to the inertial reaction force of the modal mass, $\underline{M}_r$, and these two reaction forces cancel each other. The modal damper is then reacting all of the externally applied contravariant modal force. Thus, the matrix operation sequence, $[M]^{-1}[K]\{X\}$, does not produce a state vector for just any vector, $\{X\}$, and frequency, $\omega$. It only works for the special state vectors and resonance frequencies,

$$[M]^{-1}[K]\{ \Psi \}_r = \omega_r^{\,2}\{ \Psi \}_r \qquad (5\text{-}95)$$

Or,

$$[B]\{ \Psi \}_r = \lambda_r\{ \Psi \}_r \qquad (5\text{-}96)$$

# CHAPTER VI

# EXPERIMENTAL DYNAMICAL EQUATIONS

## 6.1 The Dynamical Matrix

The undamped homgeneous equations of motion have been discussed beginning with equation (5-1). The general behaviour of a linear vibrating structure can be described by incorporating the damping matrix and a set of forcing functions:

$$[M]\{\ddot{X}(t)\}+[C]\{\dot{X}(t)\}+[K]\{X(t)\}=\{f(t)\} \tag{6-1}$$

These equations can be Fourier transformed as before [see equations (5-2), (5-3) and (5-4)], to obtain the frequency domain form,

$$\left[-\omega^2[\mathbf{M}]+\mathbf{i}\omega[\mathbf{C}]+[\mathbf{K}]\right]\{\mathbf{X}(\omega)\}=\{\mathbf{F}(\omega)\} \tag{6-2}$$

The dynamical matrix, [D], is defined as

$$[D]=\left[-\omega^2[M]+i\omega[C]+[K]\right] \tag{6-3}$$

Using the dynamical matrix, [D], allows equation (6-2) to be written more compactly as

$$[\mathbf{D}]\{\mathbf{X}(\omega)\} = \{\mathbf{F}(\omega)\} \tag{6-4}$$

The dynamical matrix actually completely characterizes the structure. All information is contained in the dynamical matrix that would be needed to predict the response of the structure at any location in any direction due to any arbitrary set of forcing functions.

Consider the task of measuring the individual elements of the dynamical matrix for a structure such as the cantelever beam shown in figure 6.1. A characterization of the beam will be approximated using six nodes. Only the vertical shear deflections will be of interest for this discussion. Accordingly the physical coordinates will be identified as $X_1$, $X_2$, $X_3$, ... , $X_6$. A set of external forcing functions at the six nodes are identified as $f_1$, $f_2$, $f_3$, ... , $f_6$.

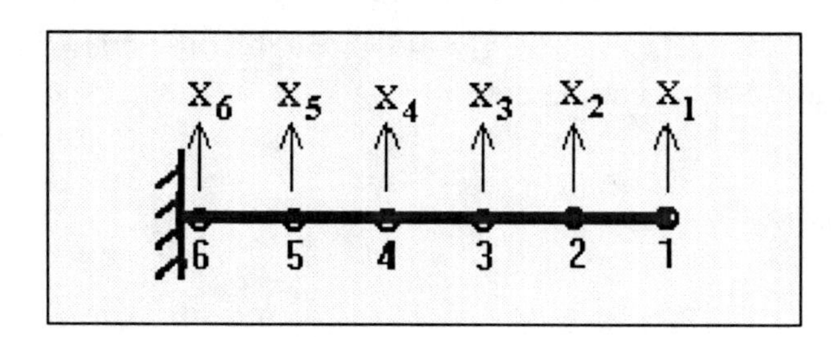


**Figure 6.1 Cantelever beam structure. Structural deformations and forces are approximated using just six nodes and corresponding physical coordinates, $X_1$, $X_2$, ... , $X_6$.**

The displacement functions are related to the forces through the dynamical matrix in the frequency domain as indicated by equation (6-4) and given explicitly for the cantelever beam six degrees of freedom as follows:

$$\begin{bmatrix} d_{11}(\omega) & d_{12}(\omega) & d_{13}(\omega) & \cdots & d_{16}(\omega) \\ d_{21}(\omega) & d_{22}(\omega) & d_{23}(\omega) & \cdots & d_{26}(\omega) \\ \vdots & \vdots & \vdots & \vdots & \vdots \\ d_{61}(\omega) & d_{62}(\omega) & d_{63}(\omega) & \cdots & d_{66}(\omega) \end{bmatrix} \begin{Bmatrix} X_1(\omega) \\ X_2(\omega) \\ \vdots \\ X_6(\omega) \end{Bmatrix} = \begin{Bmatrix} f_1(\omega) \\ f_2(\omega) \\ \vdots \\ f_6(\omega) \end{Bmatrix} \qquad (6\text{-}5)$$

Figure 6.2 sketches a hypothetical test lab setup designed to allow the measurement of just one column of the dynamical matrix, column number three. All degrees of freedom are restrained at zero displacement while enforcing a unit displacement at node number three in physical coordinate, $X_3$. The deflection is enforced using a reaction force to ground through a force transducer so that the applied force, $f_3$, is measured as well as displacement, $X_3$. The remaining nodes are restrained to ground through force transducers so that all reaction forces can be measured.

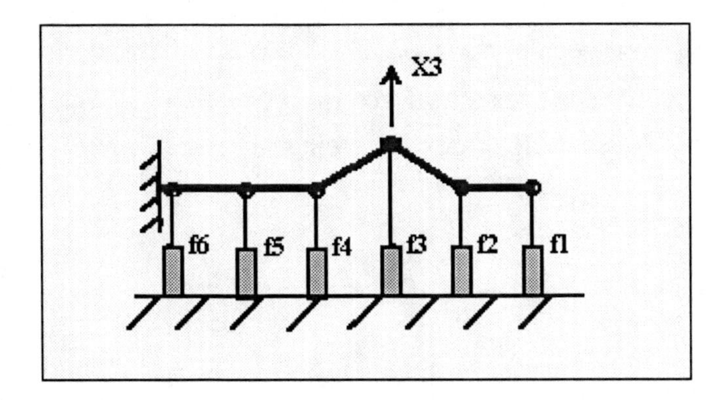


**Figure 6.2 Enforced deflection on node 3. All other nodes are restrained to zero deflection. All reaction forces, f1,f2, ..., f6, are measured with force transducers.**

The calculation of one element of the dynamical matrix, $d_{23}(\omega)$, for example, can be seen by substituting zero into all displacements except for the unit displacement at node number three.

$$\begin{bmatrix} d_{11}(\omega) & d_{12}(\omega) & d_{13}(\omega) & \cdots & d_{16}(\omega) \\ d_{21}(\omega) & d_{22}(\omega) & d_{23}(\omega) & \cdots & d_{26}(\omega) \\ d_{31}(\omega) & d_{32}(\omega) & d_{33}(\omega) & \cdots & d_{36}(\omega) \\ \vdots & \vdots & \vdots & \cdots & \vdots \\ d_{61}(\omega) & d_{62}(\omega) & d_{63}(\omega) & \cdots & d_{66}(\omega) \end{bmatrix} \begin{Bmatrix} 0 \\ 0 \\ X_3(\omega) \\ \vdots \\ 0 \end{Bmatrix} = \begin{Bmatrix} f_1(\omega) \\ f_2(\omega) \\ f_3(\omega) \\ \vdots \\ f_6(\omega) \end{Bmatrix} \tag{6-6}$$

Multiplying the second row of the dynamical matrix times the displacement column vector produces a result equal to force, $f_2(\omega)$, on the right hand side of (6-6).

$$d_{23}(\omega) \bullet X_3(\omega) = f_2(\omega) \qquad 6\text{-}7)$$

So, $d_{23}(\omega)$ is seen to correspond to the ratio, $f_2(\omega)$ divided by $X_3(\omega)$ where $f_2(\omega)$ is the reaction force (as a function of frequency) at node number 2 resulting from enforcing a displacement (function of frequency) at node number 3.

$$d_{23}(\omega) = \frac{f_2(\omega)}{X_3(\omega)} \qquad (6\text{-}8)$$

Any general element of the dynamical matrix, $d_{ji}$, corresponds to the ratio, $f_j/X_i$, where $f_j$ is the reaction force at node j caused by enforcing a displacement, $X_i$, at node i while restraining all other nodes to zero displacement.

$$d_{ji}(\omega) = \frac{f_j(\omega)}{X_i(\omega)} \qquad (6\text{-}9)$$

Notice that enforcing a displacement at just one node allows computation of one complete column of the dynamical matrix. Performing measurements for computation of the entire dynamical matrix would require enforcing displacements at each node, one node at a time, while measuring a complete set of forces each time.

It might seem attractive to check the correctness of finite element models by measuring the dynamical matrix as described above. Unfortunately there are a number of factors that make the measurement of the dynamical matrix quite impractical:

1. For most structures of interest it is not practical to build a test fixture with sufficiently high stiffness to enforce significant deflection in just one degree of freedom while restraining all other degrees of freedom.

2. The geometry of most real structures renders most of the node locations inaccessable to test fixture attachment points. There would be no way to attach a stiff connection from ground to enclosed locations.

3. The number of measurements required for moderately complex structures makes it impractical. A structure with 300 node locations times 3 directions would require enforcing displacements in 900 degrees of freedom. Reaction forces for all 900 degrees of freedom would be measured for each of the 900 displacement tests (360,000 measurements). Each displacement of force measurement would involve a collection of perhaps 1000 data points (not counting ensemble averaging).

It is for these reasons that measurement of the dynamical matrix is never attempted for structures of any complexity. Instead, the inverse of the dynamical matrix is measured, at least partially, and this inverse matrix is known as the matrix of Frequency Response Functions (FRF's).

## 6.2 The FRF Matrix

Test laboratories have for some time found it reasonably easy to apply an excitation force to a structure at a single point while measuring response motion (usually acceleration) over the entire structure. This set of measurements allows computation of one column of the matrix of Frequency Response Functions (FRF's). The full matrix of FRF's will be represented as $[\ H(\omega)\ ]$, and this is the inverse of the dynamical matrix.

$$[\ H(\omega)\ ] = [\ D(\omega)\ ]^{-1} \qquad (6\text{-}10)$$

It is almost impossible to measure displacements over an entire structure of any complexity, but commercially available accelerometers provide the relatively easy measurement of acceleration. Because of the equation (6-10) relationship, the displacement response form of the FRF will be used here, even though acceleration is the quantity actually measured in the lab. Converting data between displacement and acceleration, i.e., double integration or double differentiation, is performed easily in the frequency domain, given the simple algebraic relationship between displacement and acceleration:

$$X(\omega) = -\omega^2 A(\omega) \qquad (6\text{-}11)$$

An individual element of the $[\ H(\omega)\ ]$ matrix such as $h_{jk}(\omega)$ corresponds to the ratio, displacement $X_j$ divided by force $f_k$.

$$\mathbf{h}_{\mathbf{jk}}(\omega) = \frac{\mathbf{X}_{\mathbf{j}}(\omega)}{\mathbf{f}_{\mathbf{k}}(\omega)} \qquad (6\text{-}12)$$

The displacement measured at degree of freedom, j, is caused by a single external force applied at the degree of freedom, k. $X_j(\omega)$ and $f_k(\omega)$ are of course Fourier transforms of the time series of measured displacement and force. Figure 6.3 depicts the measurement of displacement response over all nodes of the cantelver beam due to an excitation force at node number 2. This allows computation of the FRF's making up the second column of the FRF matrix.

A full matrix of FRF's could be computed, one column at a time, by moving the excitation force from one node point to another, each time measuring the response motion at all nodes. The full matrix is represented as follows:

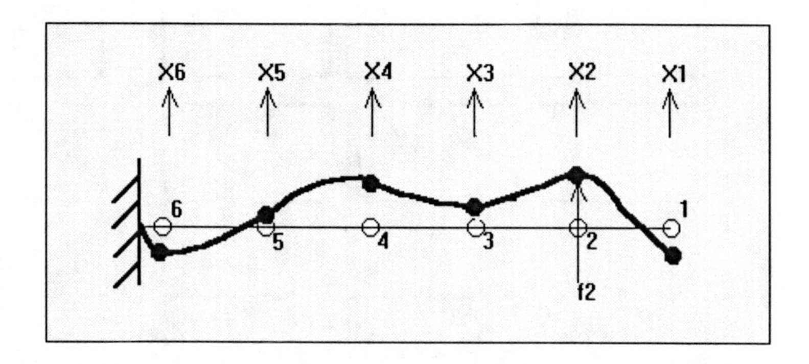


**Figure 6.3 Snaphot photo of a vibrating cantelever beam at an instant of time. The structure is excited by a dynamic forcing function applied at node number 2. Response displacements can be measured at the six nodes for physical coordinates, $X_1$, $X_2$, $X_3$, ... , $X_6$. This allows computation of the second column of the square 6x6 FRF matrix.**

$$[H(\omega)] = \begin{bmatrix} h_{11} & h_{12} & h_{13} & \cdots & h_{16} \\ h_{21} & h_{22} & h_{23} & \cdots & h_{26} \\ \vdots & \vdots & \vdots & \cdots & \vdots \\ h_{61} & h_{62} & h_{63} & \cdots & h_{66} \end{bmatrix} \tag{6-13}$$

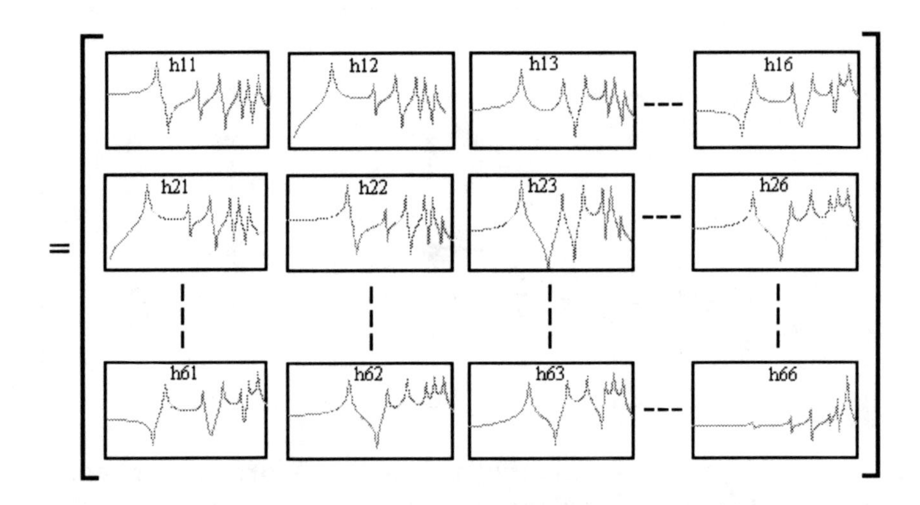


Individual FRF (Frequency Response Function) plots have been inserted into the matrix element positions of the FRF matrix to emphasize the content of each element. That is to say, each element of the FRF matrix consists of an individual Frequency Response Function of frequency, i.e., $h_{jk}(\omega) = X_j(\omega)/F_k(\omega)$ . Numerically, an individual FRF is stored as an array of complex numbers. The actual array elements of an individual FRF are typically stored as a pair of real and imaginary numbers for each element. A typical FRF may store 2048 complex number elements in a single FRF array. In this case, then, one element of the $H(\omega)$ matrix would actually consist of an array of 2048 complex numbers, one complex number for each discrete frequency of the FRF spectrum.

A test of a large aerospace structure might require 300 to 1000 or more FRF's in a single column of the FRF matrix, $[\ H(\omega)\ ]$. The number of columns would usually not exceed four to six, resulting in a rectangular matrix. Experimental FRF matrices, $H(\omega)$, are almost never square.

The FRF data appearing in the above matrix element positions are plotted as log(magnitude) vs. log(frequency). Since the numbers in an FRF array are complex, i.e., real and imaginary, the data can be plotted in many different forms. Plot forms commonly used in experimental modal analysis include the bode plot (FRF log magnitude vs. log frequency along with linear phase angle vs. log frequency), linear FRF real part vs. frequency, linear FRF imaginary part vs. frequency, overlay of real and imaginary and the Nyquist plot (linear FRF imaginary part vs. linear FRF real part). Example FRF data, generated analytically, illustrate the appearance of these types of plots, figures 6-4 thru 6-8 .

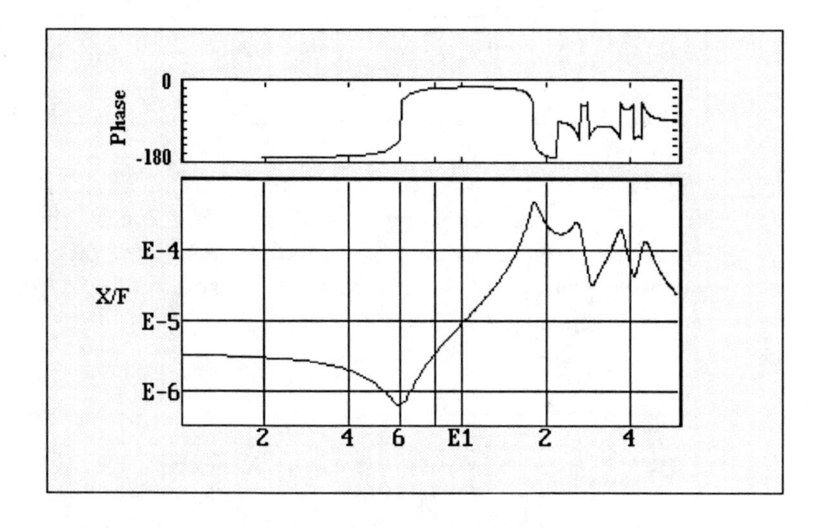


**Figure 6-4 FRF Bode plot example. Log of FRF magnitude versus log frequency (lower segment) and linear phase versus log frequency (upper segment). Peaks indicate resonance frequencies.**

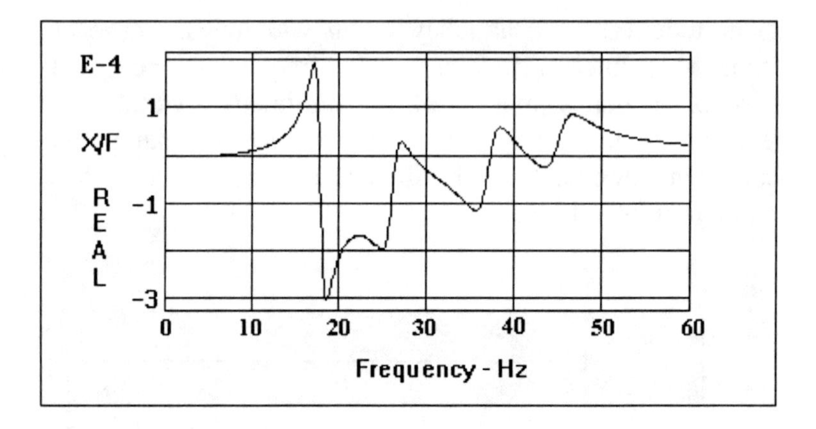


**Figure 6-5 Example plot of the real part of an FRF versus frequency. A characteristic of this function is that each resonance frequency is sandwiched by a pair of positive and negative peaks. This holds true if the resonance frequencies are well separated.**

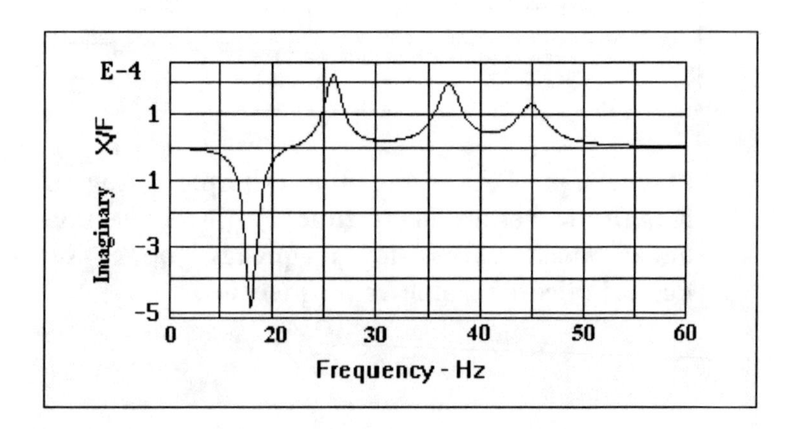


**Figure 6-6 Example plot of the imaginary part of an FRF versus frequency. Notice that it is the imaginary part that peaks positive or negative at the resonance frequencies. When the frequencies are well separated, the peak FRF values may be taken as mode coefficients. This will be illustrated in more detail later.**

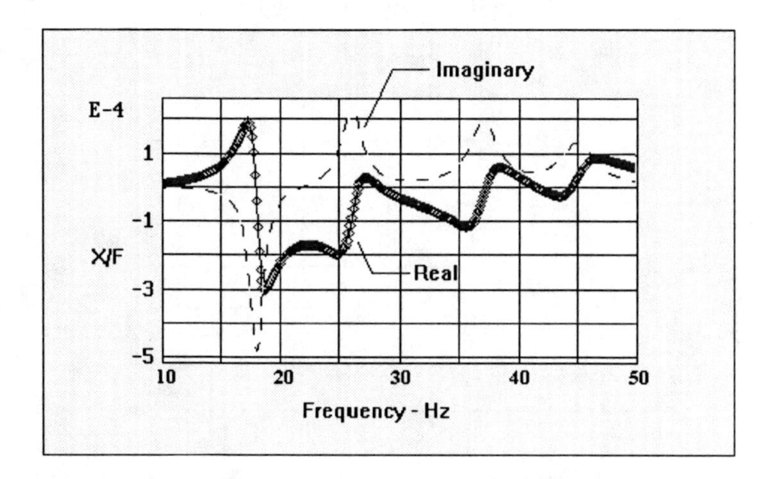


**Figure 6-7 Overlay of FRF real part and imaginary part. The imaginary positive/negative peaks at the resonance frequencies always appear midway between real positive/ negative peak pairs when resonance frequencies are well separated. This feature makes the real and imaginary overlay one of the most useful plot forms for assessing test data.**

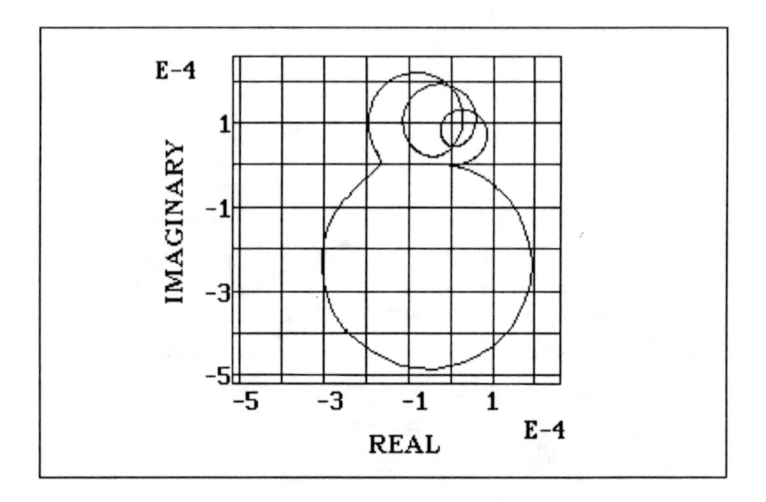


**Figure 6-8 A Nyquist plot of a Frequency Response Function. The FRF imaginary part is plotted against the real part. It is also known as an Argand plot. Later the curve will be seen as a superposition of single degree of freedom circles.**

## 6.3 Modal Frequency Response Functions

One of the primary goals of this text is to develop concepts leading to the understanding of the forced vibration of structures from the standpoint of modal coordinates. Up to this point the concept of contravariant modal forces producing modal displacements has been developed along with the relationship of modal coordinates to physical coordinates. And now, having worked through frequency response function concepts, it is appropriate to consider the payoffs that accrue when applying the modal coordinate representation to FRF's (modal frequency response functions). We have actually developed the basic mathematical machinery for writing out the relationship between the modal FRF matrix and the physical FRF matrix since the FRF is simply an operator. It is the inverse operator of the dynamical matrix operator as expressed in equation (6-10), repeated below.

$$[\, \mathbf{H}(\omega) \,] = [\, \mathbf{D}(\omega) \,]^{-1} \tag{6-14}$$

Clearly, like any operator, the $[\ H\ ]$ matrix may be expressed in modal coordinates, and the transformation from physical coordinates is developed as follows. Starting with the operator transformation from physical coordinates to modal coordinates for the dynamical matrix,

$$\begin{bmatrix} \ddots & & \\ & \underline{\mathbf{D}'} & \\ & & \ddots \end{bmatrix} = [\Psi]^T [\mathbf{D}][\Psi] \tag{6-15}$$

Taking the inverse of both sides of (6-15),

$$\begin{bmatrix} \ddots & & \\ & \underline{\mathbf{D}'} & \\ & & \ddots \end{bmatrix}^{-1} = [\Psi]^{-1} [\mathbf{D}]^{-1} [\Psi]^{-T} \tag{6-16}$$

or,

$$\begin{bmatrix} \ddots & & \\ & \underline{\mathbf{H}} & \\ & & \ddots \end{bmatrix} = [\Psi]^{-1} [\mathbf{H}][\Psi]^{-T} \tag{6-17}$$

Note that we will drop the prime on the modal FRF 's. The modal coordinate *inverse* operation of carrying *contravariant* modal vector components into *covariant* modal vector components will always be implied (recall the direct operation from Figure 4-24). The FRF matrix operators perform as follows for physical and modal coordinates:

$$\{ \mathbf{X}(\omega) \} = [ \mathbf{H}(\omega) ]\{ \mathbf{F}(\omega) \} \tag{6-18}$$

$$\{ \underline{\mathbf{X}}(\omega) \} = [ \underline{\mathbf{H}}(\omega) ]\{ \underline{\mathbf{F}}'(\omega) \} \tag{6-19}$$

where $[ \underline{\mathbf{H}} ]$ will always be diagonal.

One specific objective served by developing these operator relationships is the objective of developing a mathematical expression for a measured physical coordinate FRF as a linear superposition of modal FRF's. And this result immediately falls out of the coordinate transformation between physical and modal coordinates for the FRF matrix as indicated by equation (6-17). The $[ H ]$ matrix of FRF's may be expressed as a superposition of the modal FRF's by premultiplying and post multiplying (6-17) by $[ \Psi ]$ and $[ \Psi ]^{\mathbf{T}}$, respectively:

$$[ \mathbf{H}(\omega) ] = [ \Psi ][ \underline{\mathbf{H}}(\omega) ][ \Psi ]^{\mathbf{T}} \tag{6-20}$$

This line of thought will be pursued further a little later. The actual formula for an individual physical FRF, $h_{jk}(\omega)$, will be extracted from (6-20), but first the explicit formula for an individual modal FRF, $\underline{h}_{rr}(\omega)$, will be developed starting with the differential equations of motion.

Repeating the differential equations of motion in the frequency domain for a linear system (physical coordinates) from equation (6-2):

$$[ -\omega^{2}[ \mathbf{M} ] + i\omega[ \mathbf{C} ] + [ \mathbf{K} ] ]\{ \mathbf{X}(\omega) \} = \{ \mathbf{F}(\omega) \} \tag{6-21}$$

The differential equations of motion could be written immediately in modal coordinates without further derivation. It is reinforcing to carry out the coordinate transformation on (6-21) to obtain the same result. So, premultiply both sides of (6-21) by $[\,\Psi\,]^T$ while also sandwiching the identy, $[\Psi][\Psi]^{-1}$, just in front of the {X} column.

$$[\Psi]^T[\,-\omega^2[M] + i\omega[C] + [K]\,][\Psi][\Psi]^{-1}\{X\} = [\Psi]^T\{F\} \qquad (6\text{-}22)$$

Recall the coordinate transformation of equation (4-18), from physical {X} to modal $\{\underline{X}\}$, and the transformation from physical force {F} to contravariant modal force, $\{\underline{F}'\}$, from equation (4-40). These transformations are repeated here.

$$\{\,\underline{X}\,\} = [\,\Psi\,]^{-1}\{\,X\,\} \qquad (6\text{-}23)$$

$$\{\,\underline{F}'\,\} = [\,\Psi\,]^T\{\,F\,\} \qquad (6\text{-}24)$$

After making the substitutions using (6-23) and (6-24) it is noticed that each of the operators, [ M ], [ C ] and [ K ], are premultiplied by $[\,\Psi\,]^T$ and post multiplied by $[\Psi]$. This pre and post multiplication of course transforms each of the three operators into modal coordinates as seen earlier in equations (4-75), (4-79) and (4-81). Equation (6-22) becomes

$$[\,-\omega^2[\,\underline{M}\,] + i\omega[\,\underline{C}\,] + [\,\underline{K}\,]\,]\{\,\underline{X}(\omega)\,\} = \{\,\underline{F}'(\omega)\,\} \qquad (6\text{-}25)$$

where all three operators are of course diagonal. The matrix factor including the three operators in the summation is an expanded form of the dynamical matrix in modal coordinates, encountered earlier in equation (6-15).

Equation (6-25) represents a set of completely uncoupled second order constant coefficients differential equations (converted from a set of second order D.E.'s to a set of simple algebraic equations through use of the Fourier transform). You can get one equation by selecting any one row, say the rth row, of the operator matrices and then carry out the operator row times {X} column multiplication, setting that sum of products (one product for each operator) equal to the corresponding rth row (element) of the contravariant modal force vector on the right side of (6-25). Selecting the rth row of the operators and carrying out this row times column operation gives the rth equation,

$$(-\omega^2\underline{\mathbf{m}}_r + i\omega\underline{\mathbf{c}}_r + \underline{\mathbf{k}}_r)\underline{\mathbf{x}}_r(\omega) = \underline{\mathbf{f'}}_r(\omega) \tag{6-26}$$

Thus, a very complicated set of differential equations, equation (6-1), has been reduced to the problem of coping with just one representative algebraic equation representing a single degree of freedom. Indeed, if one row of the operator matrices of the corresponding physical coordinate algebraic equations of (6-21) had been selected for performing a similar row times {X} column multiplication, all of the physical coordinates, $x_1, x_2, x_3, \ldots$, would show up in one messy equation.

The expression for a modal FRF for any rth mode can be obtained by the appropriate division on equation (6-26):

$$\underline{\mathbf{h}}_r(\omega) = \underline{\mathbf{x}}_r(\omega)/\underline{\mathbf{f'}}_r(\omega) \tag{6-27}$$

$$\mathbf{h}_r(\omega) = 1/(-\omega^2\underline{\mathbf{m}}_r + i\omega\underline{\mathbf{c}}_r + \underline{\mathbf{k}}_r) \tag{6-28}$$

The SDOF (Single Degree Of Freedom) FRF formula of equation (6-28) will be easier to interpret after making a couple of substitutions and rationalizing the denominator (getting the imaginary number, i, out of the denominator). In order to identify the substitutions it is a good idea to back up a little and solve the homogeneous form of (6-26) for frequency. The results of this effort will also fit directly into the solution of the time domain modal differential equation. Although it may seem to be a lengthy diversion, the connecting of these two methods (frequency and time domain) should prove worth it.

## 6.4 Differential Equations In Modal Coordinates

We will continue working in the frequency domain for a while, then start again in the time domain with the second order differential equation for any rth mode of a vibrating structure. The focus now will be on the homogeneous equation.

Rewriting (6-26), setting the excitation force to zero (free vibration),

$$(-\omega^2\underline{\underline{m}}_r + i\omega\underline{c}_r + \underline{k}_r)\underline{x}_r(\omega) = 0 \qquad (6\text{-}29)$$

Nontrivial solutions for (6-29) satisfy

$$-\omega^2\underline{\underline{m}}_r + i\omega\underline{c}_r + \underline{k}_r = 0 \qquad (6\text{-}30)$$

The form is slightly altered by dividing through by $-\underline{\underline{m}}_r$ .

$$\omega^2 - i\omega(\underline{c}_r/\underline{\underline{m}}_r) - (\underline{k}_r/\underline{\underline{m}}_r) = 0 \qquad (6\text{-}31)$$

This is of course a standard quadratic equation having a solution given by the well known quadratic formula from algebra:

$$\omega = i\frac{\underline{c}_r}{2\underline{\underline{m}}_r} \pm \sqrt{-\left(\frac{\underline{c}_r}{2\underline{\underline{m}}_r}\right)^2 + \frac{\underline{k}_r}{\underline{\underline{m}}_r}} \qquad (6\text{-}32)$$

Consider the case where the damping constant is zero, i.e., $\underline{c}_r = 0$. Then $\omega$ is the undamped resonance frequency,

$$\omega_r = \sqrt{\frac{\underline{k}_r}{\underline{m}_r}} \tag{6-33}$$

A mode will be critically damped when the radical term of (6-32) is equal to zero. In this case the damping constant has a value known as the critical damping, $\underline{c}_{cr}$ .

$$\sqrt{-\left(\frac{\underline{c}_{cr}}{2\underline{m}_r}\right)^2 + \frac{\underline{k}_r}{\underline{m}_r}} = 0 \tag{6-34}$$

or,

$$\frac{\underline{c}_{cr}}{\underline{m}_r} = 2\sqrt{\frac{\underline{k}_r}{\underline{m}_r}} \tag{6-35}$$

Thus, the critical damping value is

$$\underline{c}_{cr} = 2\sqrt{\underline{k}_r \underline{m}_r} \tag{6-36}$$

Subsequent algebraic expressions will be simplified by defining two ratios. The fraction of critical damping, typically written as a decimal value or referred to as a percent, is known as the damping factor, $\zeta$:

$$\zeta_r = c_r / c_{cr} \tag{6-37}$$

or

$$\zeta_r = \frac{\underline{c}_r}{2\sqrt{\underline{k}_r \underline{m}_r}} \tag{6-38}$$

The other definition is the frequency ratio, $\beta$ .

$$\beta r = \omega / \omega r \tag{6-39}$$

Now, the frequency solution for the rth mode, equation (6-32), will be rewritten. But first, incorporate the new definitions into the $\underline{c}_r/2\underline{m}_r$ term. Multiply both sides of (6-38) by the square root of $\underline{k}_r/\underline{m}_r$ .

$$\sqrt{\frac{\underline{k}_r}{\underline{m}_r}}\zeta_r = \frac{\underline{c}_r}{2\sqrt{\underline{k}_r \underline{m}_r}}\sqrt{\frac{\underline{k}_r}{\underline{m}_r}} \qquad (6\text{-}40)$$

Combining the $\underline{m}_r$ factors and cancelling the $\underline{k}_r$ factors on the right while recognizing the expression for resonance frequency on the left results in

$$\omega_r \zeta_r = \frac{\underline{c}_r}{2\underline{m}_r} \qquad (6\text{-}41)$$

Using this sustitution and again recognizing the expression for resonance frequency, equation (6-32) may be written as

$$\omega = i\omega_r \zeta_r \pm \sqrt{-\omega_r^2 \zeta_r^2 + \omega_r^2} \qquad (6\text{-}42)$$

Factoring $\omega_r^2$ out of the radical,

$$\omega = i\omega_r \zeta_r \pm \omega_r \sqrt{1 - \zeta_r^2} \qquad (6\text{-}43)$$

The system is overdamped if $\zeta_r$ is greater than 1.0. This means that the system, if deformed from a rest position, then released, will return back to the rest position without going into free oscillations. This text will only be concerned with structures capable of vibrating freely, where $\zeta_r$ is less than 1.0.

The frequency domain second order differential equation solution has poven to be easy, quick and efficient. However, one should not lose touch with the traditional time dependent solution. We begin again with the homogenious equation,

$$\underline{\underline{m}}_r \frac{d^2}{dt^2}\underline{x}_r(t) + \underline{\underline{c}}_r \frac{d}{dt}\underline{x}_r(t) + \underline{\underline{k}}_r \underline{x}_r(t) = 0 \tag{6-44}$$

It is customary to assume a solution of the form,

$$\underline{x}_r(t) = Ae^{i\omega t} + Be^{-i\omega t} \tag{6-45}$$

where the constants A and B are determined by the initial conditions of displacement and velocity at time equals zero.

The solution could be written with a single term, still using two constants, an amplitude, C, and a phase angle, $\theta$:

$$\underline{x}_r(t) = Ce^{i(\omega t+\theta)} \tag{6-46}$$

Using (6-43) for $\omega$ and considering only the underdamped case,

$$\underline{x}_r(t) = Ce^{-\omega_r \zeta_r t} e^{i(\omega_r \sqrt{1-\zeta_r^2}\, t+\theta)} \tag{6-47}$$

To see the damped sine form of this equation, apply Euler's theorem,

$$e^{i\alpha} = \cos(\alpha) + i\sin(\alpha) \tag{6-48}$$

This theorem is proven by expanding both sides of (6-48) in a Taylor expansion and observing that the two sides match, term by term. Now, substituting (6-48) into (6-47),

$$\underline{x}_r(t) = Ce^{-\omega_r \zeta_r t}[\cos(\omega_r\sqrt{1-\zeta_r^2}\, t+\theta) + i\sin(\omega_r\sqrt{1-\zeta_r^2}\, t+\theta)] \tag{6-49}$$

Using $\omega_{dr}$ for the damped resonance frequency of the rth mode of vibration,

$$\omega_{dr} = \omega_r \sqrt{1-\zeta_r{}^2} \qquad (6\text{-}50)$$

And (6-49) is written with a little less notation,

$$\underline{\mathbf{x}}_r(t) = Ce^{-\omega_r \zeta_r t}[\cos(\omega_{dr}t + \theta) + i\sin(\omega_{dr}t + \theta)] \qquad (6\text{-}51)$$

There is set of boundary conditions in modal analysis that arises quite often in hammer impact modal testing. A hammer is selected having appropriate hammer head mass and contact tip stiffness so as to produce a very short half-sine acceleration pulse when impacting on a test structure. If the pulse duration is very short compared to the time period of one cycle of vibration at resonance, then the boundary conditions approximate an initial displacement of zero and some nonzero velocity, $v_0$. The initial velocity is of course a result of the finite duration acceleration and is equal to the area under the acceleration half-sine function, which lasts for just one half of a cycle of the resonance frequency of the hammer mass/spring component (a much higher resonance frequency than the structure's resonance frequency). Applying this boundary condition to equation (6-51), the sine part of the solution is taken with phase angle, θ, equal to zero.

$$\underline{\mathbf{x}}_r(t) = Ce^{-\omega_r \zeta_r t}[\, i\sin(\omega_{dr}t)\,] \qquad (6\text{-}52)$$

Taking the first derivative of (6-52) to get velocity,

$$v(t) = -\omega_r\zeta_r Ce^{-\omega_r \zeta_r t}[\, i\sin(\omega_{dr}t)\,] + Ce^{-\omega_r \zeta_r t}[\, i\omega_{dr}\cos(\omega_{dr}t)\,]$$

(6-53)

Setting velocity equal to $v_0$, and setting t=0 (sine is zero, cosine is 1 and $e^0$ is 1),

$$v_0 = C(i\omega_{dr}) \tag{6-54}$$

or,

$$C = \frac{-iv_0}{\omega_{dr}} \tag{6-55}$$

And using the (6-50) expression for $\omega_{dr}$,

$$C = \frac{-iv_0}{\omega_r\sqrt{1-\zeta_r^{\,2}}} \tag{6-56}$$

Substituting (6-55) into (6-52),

$$\underline{x}_r(t) = \frac{v_0}{\omega_{dr}} e^{-\omega_r \zeta_r t} \sin(\omega_{dr} t) \tag{6-57}$$

or, using (6-50),

$$\underline{x}_r(t) = \frac{v_0}{\omega_r\sqrt{1-\zeta_r^{\,2}}} e^{-\omega_r \zeta_r t} \sin(\omega_r\sqrt{1-\zeta_r^{\,2}}\, t) \tag{6-58}$$

The damped sine function corresponding to this solution is plotted as Figure 6-1 for values of resonance frequency equal to 10 Hz and damping factor, $\zeta_r = .04$.

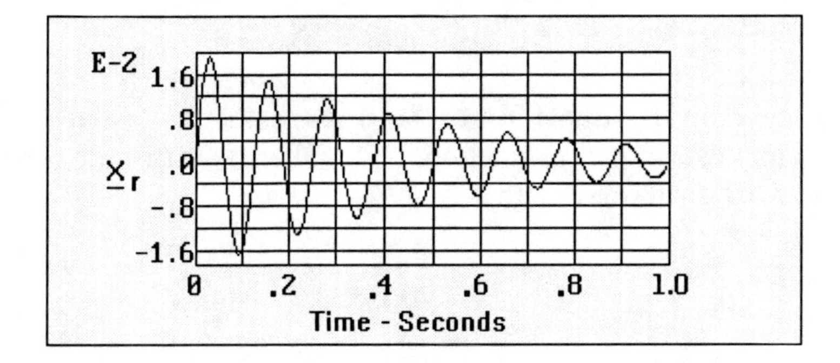


**Figure 6-9. Damped sine function solution to second order differential equation for boundary conditions of displacement equals zero at time zero and some initial velocity, $v_0$. These boundary conditions are approximated during hammer impact on a dynamic structure when the impact acceleration pulse is short compared to the time period of the structure resonance frequency.**

This impact response solution is important in experimental modal analysis for reasons other than application to hammer impact testing. The equation (6-58) solution turns out to be the inverse transform of a single degree of freedom FRF. Conversely, a Fourier transform of equation (6-58) will produce a single degree of freedom FRF in the frequency domain. This is understood intuitively by realizing that the Fourier transform of an ideal impulse function, a delta function, having a time duration that approaches zero in the limit, has a constant value across the entire frequency range. Therefore, when the FRF, h(ω), is formed from the ratio, X(ω)/F(ω), the F(ω) force spectrum in the denominator is a constant, say a, so that the FRF is

$$\mathbf{h}(\omega) = \mathbf{X}(\omega)/\mathbf{a} \tag{6-59}$$

And if the force spectrum has a constant value of 1.0, then the FRF is just the Fourier transform of the response function:

$$\mathbf{h}(\omega) = \mathbf{X}(\omega) \tag{6-60}$$

And the inverse transform of X(ω) for that case is what is known as the impulse response function, h(t). That is, performing the inverse Fourier transform of both sides of (6-60),

$$\mathcal{F}^{-1}\{\, \mathbf{h}(\omega)\,\} = \mathcal{F}^{-1}\{\, \mathbf{X}(\omega)\,\} \tag{6-61}$$

or,

$$\mathbf{h}(t) = \mathbf{x}(t) \tag{6-62}$$

The Fourier transform of the Figure 6-9 function is shown in Figure 6-10.

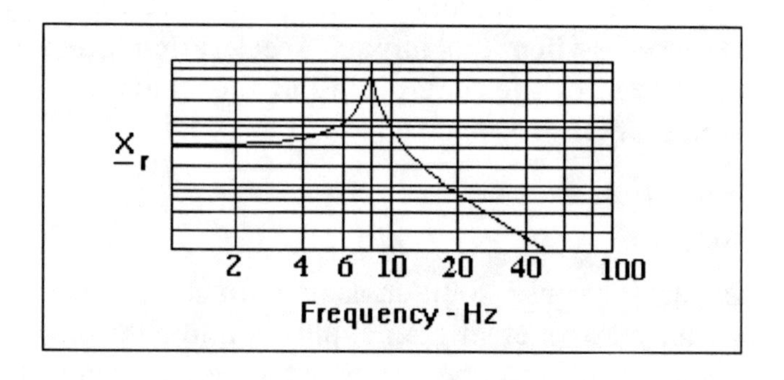


**Figure 6-10. Fourier transform of the decayed sine function of Figure 6-9. The Fourier transform of the free decay function is the same functional form as a SDOF FRF (Frequency Response Function). Figures 6-9 and 6-10 are Fourier transform pairs so that the Figure 6-9 function is an impulse response function inverse transform of Figure 6-10.**

Once the impulse response function is understood as the inverse transform of the FRF obtained from the hammer impact test, i.e., Figure 6-9 is the inverse transform of Figure 6-10, then it is quickly realized that the inverse transform of any FRF produces an impulse response function, no matter how the FRF was acquired, using random excitation or sine sweep, for example. After an FRF has been produced, it has no knowledge or memory of what method was used to compute it. Figure 6-12 is an impulse response function produced by performing an inverse transform on the FRF of Figure 6-11. There is no way of knowing what method was used to compute the FRF of Figure 6-11. Note that four resonance frequencies were present in the FRF and four decayed sine functions superimpose to form the impulse response function.

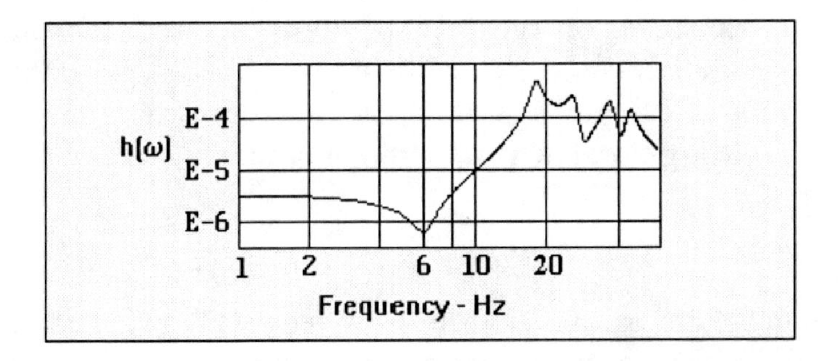


**Figure 6-11. A FRF (Frequency Response Function) with four resonance frequencies. The same FRF could have been measured using any of several different methods such as single point random, multipoint random, hammer hit, sine sweep and others. The inverse transform of this FRF is shown in Figure 6-12.**

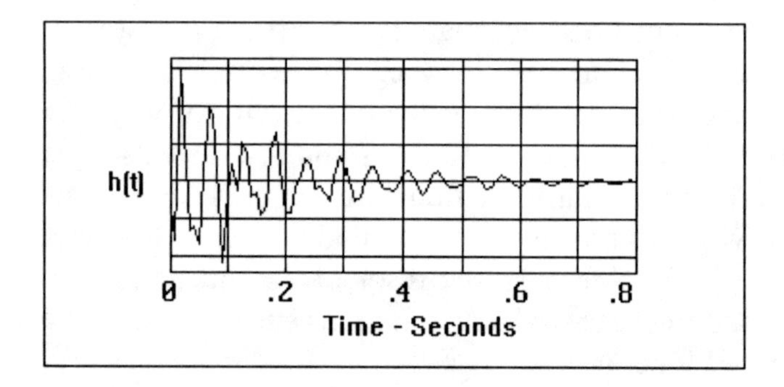


**Figure 6-12. An impulse response function, h(t). It was computed by performing an inverse Fourier transform on the FRF of Figure 6-11. This same function could be produced by applying an impact force having a very short pulse duration such as hitting the structure with a hammer.**

This transform pair relationship between FRF's and impulse response functions is the basis for some important curve fitting methods to be studied later in the text.

Before concluding this section on the subject of differential equation solutions, there is another representation that should be considered. This representation uses the Laplace transform s plane and will be discussed later in conjunction with time domain curve fitting. Equation (6-46) can be written in the form

$$\underline{x}_r(t) = Ce^{s_r t+\theta} \tag{6-63}$$

or,

$$\underline{x}_r(t) = Ce^{\sigma_r t}e^{i(\omega_{dr} t+\theta)} \tag{6-64}$$

where the Laplace parameter, $s_r$, is

$$s_r = \sigma_r + i\omega_{dr} \tag{6-65}$$

and

$$\sigma r = \omega r \zeta r \tag{6-66}$$

$$\omega_{d_r} = \omega_r \sqrt{1 - \zeta_r^{\,2}} \tag{6-67}$$

This completes the roundup of most of the formulas needed to complete the earlier discussion of FRF's in section 6.3. That discussion will continue now, picking up again with equation (6-28).

## 6.5 More On Modal FRF's

The substitutions are now available to put the modal FRF equation into a form more easily interpreted. Equation (6-28) for the modal FRF is repeated here.

$$\underline{h}_r(\omega) = \frac{1}{(-\omega^2 \underline{m}_r + i\omega \underline{c}_r + \underline{k}_r)} \tag{6-68}$$

Multiply on the outside of the denominator by the factor, $\omega_r^2 \underline{m}_r$, and divide each term inside the denominator parentheses by the same factor:

$$\underline{h}_r(\omega) = \frac{1}{\omega_r^{\,2} \underline{m}_r \left(-\frac{\omega^2}{\omega_r^{\,2}} + i\frac{\omega \underline{c}_r}{\omega_r^{\,2} \underline{m}_r} + \frac{\underline{k}_r}{\omega_r^{\,2} \underline{m}_r}\right)} \tag{6-69}$$

Group the real part and factor terms for easy recognition of substitutions to be made.

$$\underline{\mathbf{h}}_r(\omega) = \frac{1}{\omega_r^2 \underline{\mathbf{m}}_r [(1 - \frac{\omega^2}{\omega_r^2}) + 2i\frac{\omega}{\omega_r}(\frac{\underline{\mathbf{c}}_r}{2\omega_r \underline{\mathbf{m}}_r})]} \quad (6\text{-}70)$$

Replace circular resonance frequency, $\omega_r$, with square root of $\underline{k}_r/\underline{m}_r$ from equation (6-33) and further simplify (6-70).

$$\underline{\mathbf{h}}_r(\omega) = \frac{1}{\omega_r^2 \underline{\mathbf{m}}_r [(1 - \frac{\omega^2}{\omega_r^2}) + 2i\frac{\omega}{\omega_r}(\frac{\underline{\mathbf{c}}_r}{2\sqrt{\underline{\mathbf{k}}_r \underline{\mathbf{m}}_r}})]} \quad (6\text{-}71)$$

Now, use the substitutions for $\beta_r$ from (6-39) and $\zeta_r$ from (6-38).

$$\underline{\mathbf{h}}_r(\omega) = \frac{1}{\omega_r^2 \underline{\mathbf{m}}_r [(1 - \beta_r^2) + 2i\beta_r\zeta_r]} \quad (6\text{-}72)$$

Rationalize the denominator by multiplying top and bottom by the complex conjugate of the denominator complex factor as follows.

$$\underline{\mathbf{h}}_r(\omega) = \frac{1}{\omega_r^2 \underline{\mathbf{m}}_r [(1 - \beta_r^2) + 2i\beta_r\zeta_r]} \bullet \frac{[(1 - \beta_r^2) - 2i\beta_r\zeta_r]}{[(1 - \beta_r^2) - 2i\beta_r\zeta_r]}$$

(6-73)

The final form to be used for the SDOF modal FRF is

$$\underline{\mathbf{h}}_r(\omega) = \frac{(1 - \beta_r^2) - 2i\beta_r\zeta_r}{\omega_r^2 \underline{\mathbf{m}}_r [(1 - \beta_r^2)^2 + 4\beta_r^2\zeta_r^2]} \quad (6\text{-}74)$$

The SDOF (Single Degree Of Freedom) modal FRF is now in a form easily interpreted. The function is seen now as the sum of a real function and an imaginary function of frequency, where the frequency has been put into a convenient dimensionless form, $\omega/\omega_r = \beta_r$. The absolute magnitude, $|\underline{\mathbf{h}}_r(\omega)|$, is easily computed as the square root of the complex FRF times its conjugate:

$$|\underline{h}_r(\omega)| = \sqrt{\frac{(1-\beta_r^2)^2 + 4\beta_r^2\zeta_r^2}{\omega_r^4 \underline{m}_r^2[(1-\beta_r^2)^2 + 4\beta_r^2\zeta_r^2]^2}} \qquad (6\text{-}75)$$

$$|\underline{\mathbf{h}}_r(\omega)| = \frac{1}{\omega_r^2 \underline{\mathbf{m}}_r \sqrt{(1-\beta_r^2)^2 + 4\beta_r^2\zeta_r^2}} \qquad (6\text{-}76)$$

The magnitude of the modal FRF is plotted in Figure 6-13. The function is easily sketched, noticing the effect $\beta_r$ has on $|\underline{h}_r|$ near $\beta_r=0$, $\beta_r=1$ and $\beta_r$ approaching infinity. When $\beta_r=1$, i.e., $\omega=\omega_r$, for example,

$$|\underline{\mathbf{h}}_r(\omega_r)| = \frac{1}{2\omega_r^2 \underline{\mathbf{m}}_r \zeta_r} \qquad (6\text{-}77)$$

This suggests a normalization for modal mass when characterizing structures using FRF's:

$$\underline{\mathbf{m}}_r = \frac{1}{2\omega_r^2 \zeta_r |\underline{\mathbf{h}}_r(\omega_r)|} \qquad (6\text{-}78)$$

where a particular physical degree of freedom could be selected as FRF reference. This is an important detail and will soon become clear.

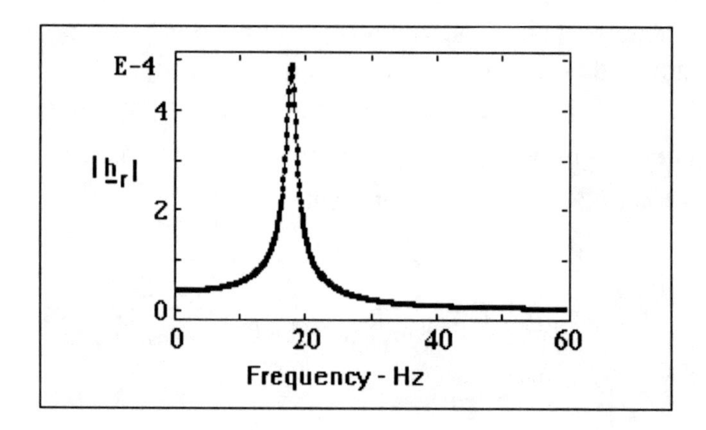


**Figure 6-13. A plot of the modal FRF magnitude versus frequency. Both ordinate and abscissa are linear. The peak at 18 Hz indicates the resonance frequency. This function is easily sketched from equation (6-76).**

The real part of the FRF is easily singled out from equation (6-74) as

$$\mathbf{R}\underline{\mathbf{h}}_r(\omega) = \frac{(1-\beta_r^{\,2})}{\omega_r^{\,2}\,\underline{\mathbf{m}}_r[(1-\beta_r^{\,2})^2 + 4\beta_r^{\,2}\zeta_r^{\,2}]} \tag{6-79}$$

At resonance the real part is zero ($\beta_r$=1). The real part is easy to evaluate at zero Hz ($\beta_r$=0).

$$\mathbf{R}\underline{\mathbf{h}}_r(0) = \frac{1}{\omega_r^{\,2}\,\underline{\mathbf{m}}_r} \tag{6-80}$$

This result has more physical significance if the formula for the circular resonance frequency is substituted into (6-80). Since $\underline{k}_r=\omega_r^2\underline{m}_r$,

$$\mathbf{R\underline{h}_r(0) = \frac{1}{\underline{k}_r}} \tag{6-81}$$

So, the modal FRF measurement at zero Hz is equivalent to a static measurement of modal flexibility or the modal influence coefficient. The matrix of modal influence coefficients is related to the matrix of static influence coefficients in physical coordinates as shown in equation (6-20) for FRF values for any frequency, including zero Hz (the static condition). Using the notation, [ G ], for the physical influence coefficient matrix and [ $\underline{G}$ ] for the modal matrix (not to be confused with the metric matrix, [ G ]),

$$[\,\mathbf{G}\,] = [\,\Psi\,][\,\underline{\mathbf{G}}\,][\,\Psi\,]^{\mathbf{T}} \tag{6-82}$$

Conceptually, this suggests that the physical coordinate stiffness matrix, [ K ], could be obtained from modal test data by taking the inverse of (6-82):

$$[\,\mathbf{K}\,] = [\,\Psi\,]^{-\mathbf{T}}[\,\underline{\mathbf{G}}\,]^{-1}[\,\Psi\,]^{-1} \tag{6-83}$$

The implications of (6-82) and (6-83) will be discussed later, but it should be noted that a very large number of modes could be required to achieve a reasonable approximation to the stiffness matrix.

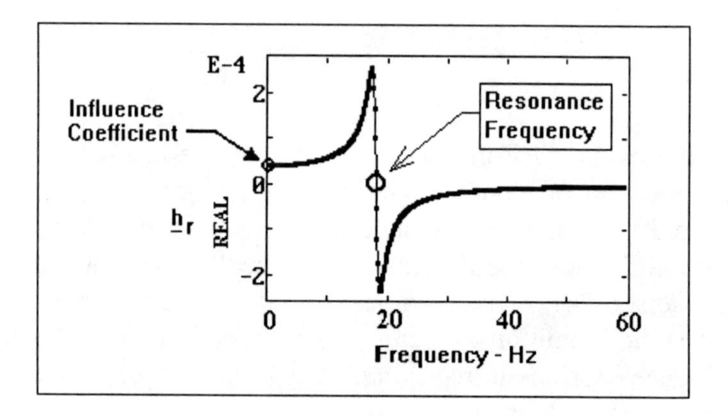


**Figure 6-14. The modal FRF real part as a function of frequency. The real function crosses zero at the resonance frequency.**

The imaginary part of the FRF, from equation (6-74), is

$$\underline{\mathbf{h}}_r(\omega) = \frac{-2i\beta_r\zeta_r}{\omega_r{}^2\,\underline{\mathbf{m}}_r[(1-\beta_r{}^2)^2 + 4\beta_r{}^2\zeta_r{}^2]} \tag{6-84}$$

This function is plotted in Figure 6-15. The resonance frequency occurs at the negative peak (valley).

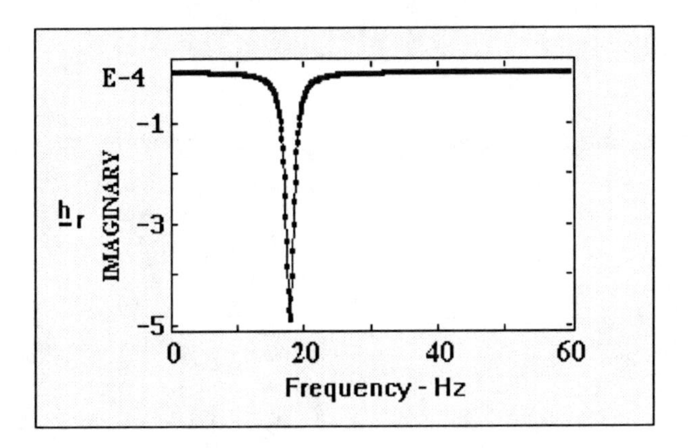


**Figure 6-15. The modal FRF imaginary part as a function of frequency. The resonance frequency occurs at the minimum value (valley).**

An overlay of the real part and imaginary part is shown as Figure 6-16. The plot windows into a frequency range around the resonance frequency. The imaginary part negative peak at resonance is sandwiched between the real part positive and negative peaks. The frequency span from the real part positive peak (located below the resonance) across the resonance to the real part negative peak (located above the resonance frequency) is called the structural bandwidth for that mode. An estimate of the damping factor can be made using the measurement of the structural bandwidth, SB, and the resonance frequency, $\omega_r$. It can be shown, as indicated in Figure 6-16, that the magnitude of the real part is equal to the magnitude of the imaginary part at the structural bandwidth frequency boundaries. This means that equation (6-79), the real part, can be set equal to equation (6-84), the real coefficient of the imaginary part, from which the structural bandwidth, SB, can be estimated. Since the denominators of (6-79) and (6-84) are the same, just the numerators are set equal:

$$1-\beta_r^2 = -2\zeta_r\beta_r \qquad (6\text{-}85)$$

$$\beta_r^2 - 2\zeta_r\beta_r - 1 = 0 \qquad (6\text{-}86)$$

Solving the quadratic equation for $\beta_r$ (ignoring negative frequency),

$$\beta_r = \zeta_r + \sqrt{\zeta_r^2 + 1} \qquad (6\text{-}87)$$

For $\zeta_r^2 << 1$,

$$\beta_r \cong 1 + \zeta_r \qquad (6\text{-}88)$$

Remembering $\beta_r = \omega/\omega_r$, the upper boundary frequency, $\omega_U$, of the structural bandwidth is

$$\omega_U \cong \omega_r(1 + \zeta_r) \qquad (6\text{-}89)$$

The structural bandwidth, SB, is equal to the difference between the upper frequency, $\omega_U$, and the lower frequency, $\omega_L$, where $\omega_L = \omega_r\,(1-\zeta_r)$. That is $SB = \omega_U - \omega_L$. Therefore,

$$\mathbf{SB} \cong (\omega_r + \zeta_r\omega_r) - (\omega_r - \zeta_r\omega_r) \qquad (6\text{-}90)$$

$$\mathbf{SB} \cong 2\zeta_r\omega_r \qquad (6\text{-}91)$$

and

$$\zeta_r \cong \frac{SB}{2\omega_r} \qquad (6\text{-}92)$$

When planning the measurement of FRF's it is useful to compute the frequency resolution, i.e., $\Delta\nu$, given an assumption about the damping factor, in order to assure that a sufficient number of discrete frequency points will be acquired across the structural bandwidth. The lowest resonance frequency usually drives this requirement. Some engineers like to see at least four or five frequency lines within the structural bandwidth. Assuming it is desired to have a frequency increment, $\Delta\omega$, that is 20% of the structural bandwidth, for example,

$$\Delta\omega = 0.2\ \mathrm{SB} \tag{6-93}$$

$$\Delta\omega = 0.4\omega_r\zeta_r \tag{6-94}$$

or, for frequency, $\nu$, in Hz,

$$\Delta\nu = 0.4\nu_r\zeta_r \tag{6-95}$$

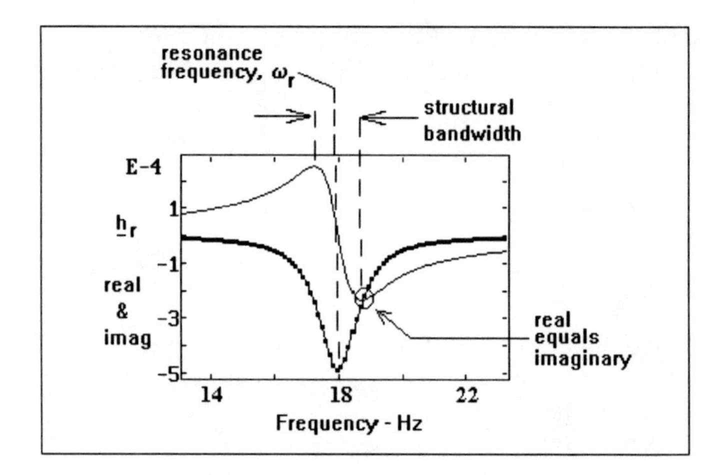


**Figure 6-16. Overlay of modal FRF real part and imaginary part. The resonance frequency occurs at the imaginary minimum and the frequency range spanned by the real part peak and valley defines the structural bandwidth. These two points on the real part have an absolute value equal to one half the absolute value of the imaginary at the structural bandwidth frequencies.**

A modal FRF Nyquist plot (Argand plane) is shown in Figure 6-17. This plots the imaginary part (ordinate) versus the real part (abscissa). The frequency is a parameter which increases along the curve in a clockwise direction around the circle. Data points corresponding to discrete sampled frequencies are displayed. The frequency samples are equally spaced in the frequency domain with a constant frequency increment, $\Delta\omega$, but increase their spacing along the curve with respect to path length, L, as the resonance frequency is approached. Then the spacing gradually decreases as frequency increases further above the resonance. A partial derivative of frequency with respect to path length along the circle would reach a minimum at the resonance frequency (path length spacing reaches a maximum). The second derivative of frequency with respect to the path length, L, would be equal to zero at this point.

$$\frac{d^2\omega}{dL^2} = 0 \qquad (6\text{-}96)$$

Equation (6-96) is sometimes used for estimating the resonance frequency when applying the circle curve fit method. This text will use the inverse, $d^2L/d\omega^2 = 0$.

The data plotted in Figure 6-17 represents the FRF equation with viscous modal damping. Actually, a true mathematical circle results only for a modal FRF that is based on structural damping. However, an FRF measured on a structure having viscous damping provides Nyquist data (Argand plane data) having a good enough approximation to a circle (particularly in the neighborhood of the resonance) to allow good estimates of modal parameters based on the use of the circle equations.

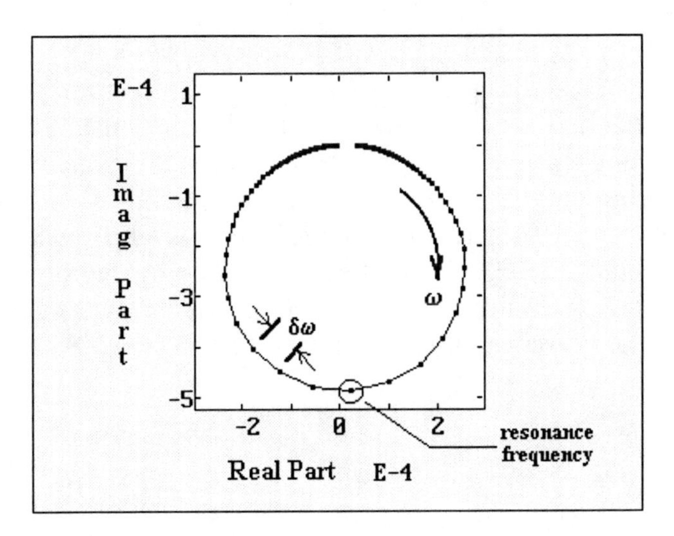


**Figure 6-17. A modal FRF Nyquist plot. The imaginary part is plotted against the real part. Frequency increases clockwise around the circle. Frequency points are evenly spaced in the frequency domain (fixed Δω), but the discrete frequency points increase their spacing along the curve as the resonance point is approached.**

Figure 6-18 summarizes the concepts we've been dealing with in this chapter: Dynamic modal forces and modal response displacement, modal FRF's and the associated modal parameters such as modal mass, modal stiffness, modal damping, modal damping factor and resonance frequency. Figure 6-18 depicts application of the modal force, $\underline{f}'_2$, for a cantelever beam. The modal force has been applied over an extended frequency range. This could have been accomplished with either a hammer impact, sine sweep or broad band random excitation. Whatever the form of the force vs. time function, the spatial distribution of forces has been maintained so as to produce a pure modal force, $\underline{f}'_2$. If this has been successful at each frequency line of the spectrum, then the transformation from physical coordinate forces to modal coordinate forces will produce zero values for every element in the modal column vector except for the modal force number two component as shown in the figure.

Further, assuming the modal FRF resulted from performing a sine sweep, it is quite remarkable that the sinusoidal forces swept right through the other resonance frequencies of the structure without peaking up at their resonances or leaving any sign of their presence. This point is often not appreciated by the student of structural dynamics. But, if a sufficient number of degrees of freedom are excited with just the right set of forces sweeping over a broad frequency range, this is exactly the physical result that could be achieved.

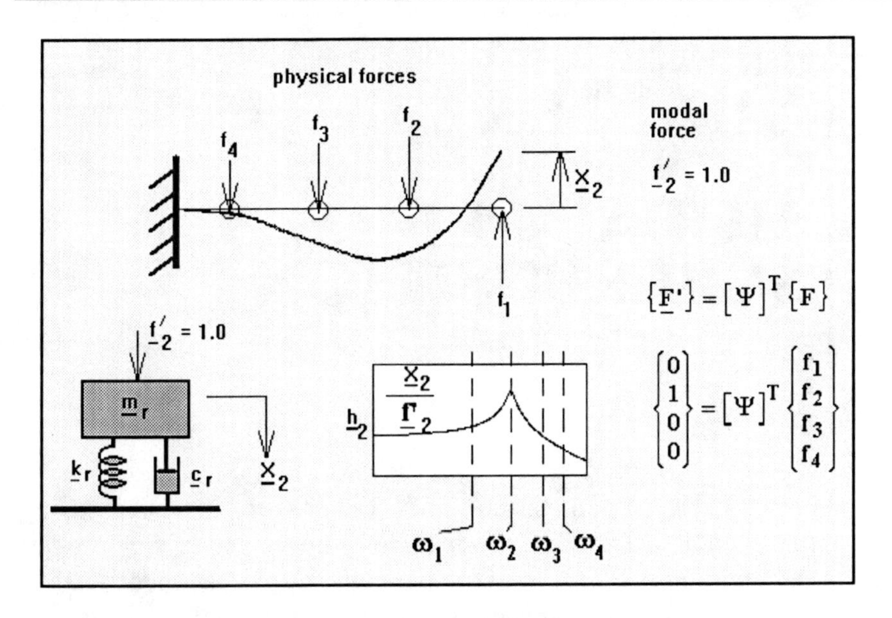


**Figure 6-18. A summary of dynamic modal force and response concepts. A contravariant modal force for the second mode of a cantelever beam has been applied over a broad frequency range. A set of physical forces has been selected so that a modal force of 1.0 will be applied obtaining displacement response in mode number two only, over the entire frequency range. A single degree of freedom modal FRF results, peaking up at the resonance frequency, $\omega_2$.**

Having developed the modal FRF concept, it is appropriate to further develop a concept that was introduced in section 6.3 with equation (6-20). This is the concept of modal FRF superposition as an analytical representation of the physical FRF.

# CHAPTER VII

# MODAL FRF SUPERPOSITION

Modal FRF superposition and the reverse, decomposing an experimentally acquired physical FRF into separate single degree of freedom modal FRFs, are fundamental concepts underlying modal analysis. The extraction of the mode shapes and associated modal parameters (resonance frequency, modal mass, modal stiffness and modal damping) from measured physical FRFs is called Modal Analysis.

A solid grasp of modal responses, motions due to applied contravariant modal forces, enables the understanding of modal FRFs and the way in which they superimpose to yield the measured physical FRF. And this understanding, in turn, provides the basis for comprehending the decomposition of measured FRFs into the separate modal FRFs. First, we consider modal forces and modal responses in a little more detail.

## 7.1 General Modal Force and Modal Response

Figure 6-10 summarized the basic dynamic modal force and response concept. This process is represented a little more generally in Figure 7-1, where each of the first four modes of a cantelever beam are represented with their equivalent modal mass, modal spring and modal damper. The picture here depicts results of four separate sine sweep tests. The set of physical forces are adjusted for each test so that only one mode is driven at a time, producing response across the entire frequency range in just that one mode. And the resulting FRF, $\underline{X}_r(\omega)/\underline{f}'_r(\omega)$, is a single degree of freedom modal FRF. The application of pure contravariant modal forces is assured by the coordinate transformation from physical to contravariant modal coordinates for the appropriate set of physical forces, as developed in equation (4-40).

$$\{\underline{F}'\} = [\Psi]^T \{F\} \tag{7-1}$$

This is a spatial relationship and holds good regardless of the time dependence of the forcing functions. In other words, it doesn't matter at what frequencies the forces are being applied. The phase angles associated with each frequency, relative to the spatial reference point, are important, because that is part of the spatial definition of the force distribution.

And once a pure contravariant modal force is applied, only the corresponding covariant modal displacement can respond, as guaranteed by the diagonal modal FRF inverse operator in equation (6-19), repeated here:

$$\begin{Bmatrix} \underline{X}_1 \\ \underline{X}_2 \\ \vdots \\ \underline{X}_r \end{Bmatrix} = \begin{bmatrix} \underline{h}_{11} & 0 & \cdots & 0 \\ 0 & \underline{h}_{22} & \cdots & 0 \\ \vdots & \vdots & \vdots & \vdots \\ 0 & 0 & \cdots & \underline{h}_{rr} \end{bmatrix} \begin{Bmatrix} \underline{f}_1' \\ \underline{f}_2' \\ \vdots \\ \underline{f}_r' \end{Bmatrix} \tag{7-2}$$

Conceptually, it would be possible to drive just one mode of a structure with random excitation. The same random forcing function would be applied at each degree of freedom (each point on the structure), but scaled by the proper factor, including sign, so as to maintain the required contravariant modal force vector at each instant of time.

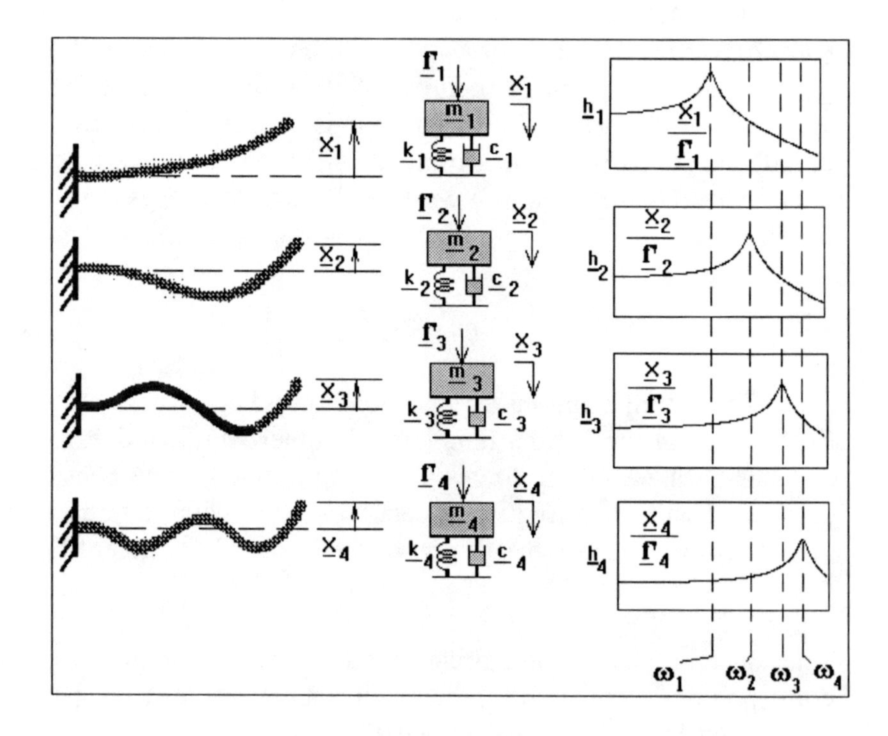


**Figure 7.1 A general picture of individual modal sine sweep tests. Four separate modal sine sweep tests are performed, each time driving a different mode of the cantelever beam. A contravariant modal force pattern is always used so that only the modal single degree of freedom response results.**

As an illustration, this process would be represented for mode number 2 of our cantelever beam example by the equations,

$$R(t)\begin{Bmatrix} 0 \\ 1 \\ 0 \\ 0 \end{Bmatrix} = [\Psi]^T \begin{Bmatrix} f_1 \\ f_2 \\ f_3 \\ f_4 \end{Bmatrix} R(t) \qquad (7\text{-}3)$$

Or, if you wish to know what set of forces are needed to produce the desired modal force, exciting mode number 2 only, just perform the reverse of the equation (7-3) computation.

$$\begin{Bmatrix} f_1 \\ f_2 \\ f_3 \\ f_4 \end{Bmatrix} = [\Psi]^{-T} \begin{Bmatrix} 0 \\ R(t) \\ 0 \\ 0 \end{Bmatrix} \tag{7-4}$$

Having an understanding of the dynamic excitation and response of individual modes, now consider the effect of applying four modal forces simultaneously, in contravariant modal force superposition.

## 7.2 Superimposing Multiple Modal Forces

If a single physical coordinate random time series force is applied at the end of a cantelever beam, as illustrated in Figure 7-2, it is equivalent to applying a superposition of four modal forces. This is implied directly from the coordinate transformation from physical coordinates to contravariant modal coordinates:

$$\begin{Bmatrix} \underline{f}'_1(t) \\ \underline{f}'_2(t) \\ \underline{f}'_3(t) \\ \underline{f}'_4(t) \end{Bmatrix} = [\Psi]^{T} \begin{Bmatrix} f_1(t) \\ 0 \\ 0 \\ 0 \end{Bmatrix} \tag{7-5}$$

The contravariant modal force amplitudes are easy to read from the row times column multiplication of (7-5).

$$\underline{f}'_1 = \Psi_{11}\ f_1(t) \tag{7-6}$$

$$\underline{f}'_2 = \Psi_{12}\ f_1(t) \tag{7-7}$$

$$\underline{f}'_3 = \Psi_{13}\ f_1(t) \tag{7-8}$$

$$\underline{f}'_4 = \Psi_{14}\ f_1(t) \tag{7-9}$$

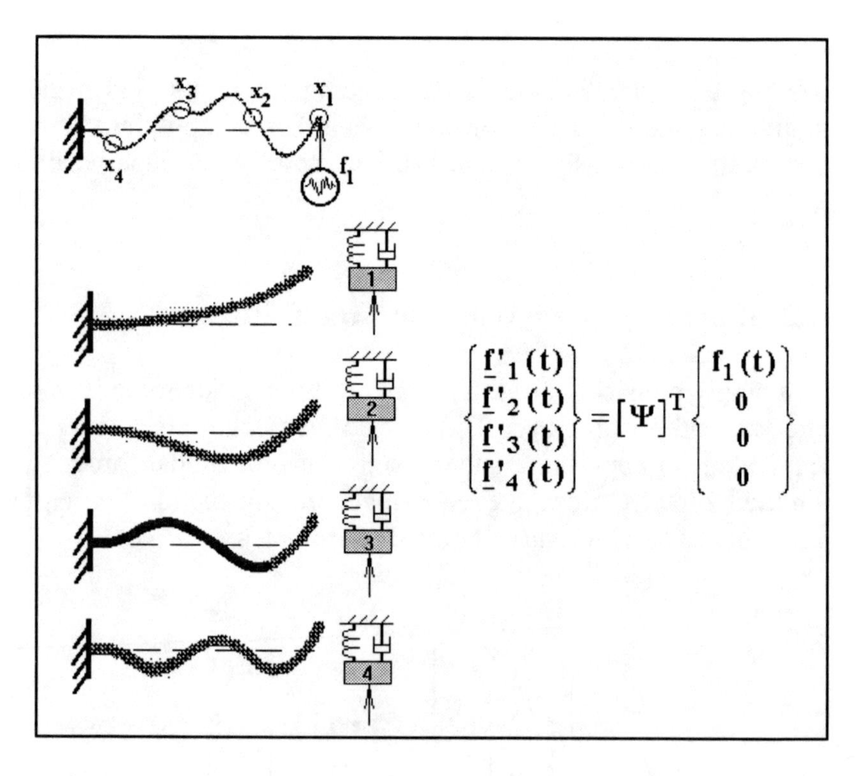


**Figure 7-2 A physical random force applied to the end of a cantelever beam is equivalent to applying a superposition of modal forces. All four modes are driven. Each mode responds over the entire frequency range of the random function.**

Thus, all four modes of the beam are driven. And if the random forcing function is broad band random, each mode is responding over the entire frequency spectrum.

If a set of random physical forces (either correlated or uncorrelated) is applied to each of the points on the beam, this set of forces is also equivalent to some superposition of random modal forces. This is represented in Figure 7-3.

Again, this result is indicated by the coordinate transformation from physical forces to contravariant modal forces:

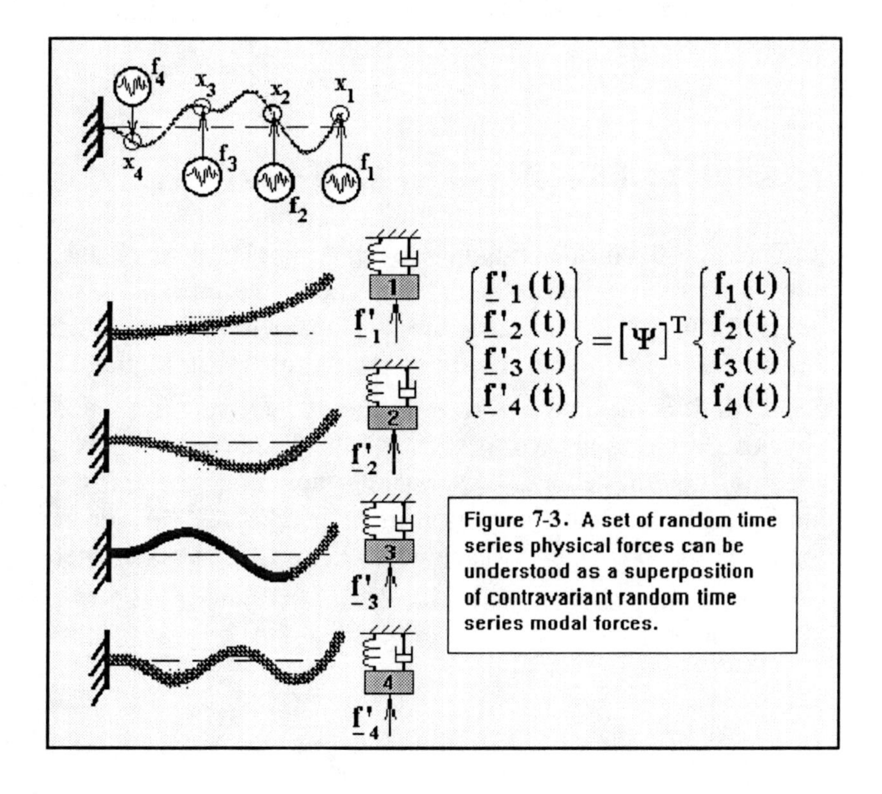


**Figure 7-3.** A set of random time series physical forces can be understood as a superposition of contravariant random time series modal forces.

$$\begin{Bmatrix} \underline{f}'_1(t) \\ \underline{f}'_2(t) \\ \underline{f}'_3(t) \\ \underline{f}'_4(t) \end{Bmatrix} = [\Psi]^T \begin{Bmatrix} f_1(t) \\ f_2(t) \\ f_3(t) \\ f_4(t) \end{Bmatrix} \tag{7-10}$$

Any one of the modal forces, $\underline{f}'_r$, is computed as

$$f'_r = \Psi_{1r} f_1(t) + \Psi_{2r} f_2(t) + \Psi_{3r} f_3(t) + \Psi_{4r} f_4(t) \tag{7-11}$$

## 7.3 Superposition Of Modal Responses

Refocus now on the example of the single broad band random physical force applied to grid point one at the end of the cantelever beam. Consider in particular the displacement response at grid point number two, $X_2$. The previous discussion directed attention to the superposition of modal forces as being equivalent to the applied physical force, as shown in Figure 7-3. The attention now is on the superposition of the modal responses. Each mode shape is responding individually, and the summation of their responses at grid point two produces the physical displacement, $X_2$. This is a statement of the coordinate transformation from modal displacements to physical displacements.

$$\{X\} = [\Psi]\{\underline{X}\} \tag{7-12}$$

For the physical displacement at grid point two, $X_2$,

$$X_2 = \sum_{r=1}^{4} \Psi_{2r} \underline{X}_r \qquad (7\text{-}13)$$

Figure 7-4 illustrates this process. The general physical beam deformation at some instant of time during vibration, as though you captured a stop action photo of the beam, is depicted in the upper left corner of Figure 7-4. Application of a random vibratory force at the end of the beam is represented as $f_1$, and the particular measured response displacement at point number two on the beam is represented as $X_2$. As in previous figures, the first four modal deformation patterns are shown with modal displacements indicated as $\underline{X}_1$, $\underline{X}_2$, $\underline{X}_3$ and $\underline{X}_4$. The amount of physical displacement at point number two on the beam, contributed by each of the modal deformations, is represented by the value obtained by scaling the modal displacement of each mode shape. The relevant mode coefficient is recognized in the coordinate transformation matrix, $[\Psi]$, used in equations (7-12) and (7-13) above.

Thus, the individual modal deformation contributions to the physical displacement at point two on the beam are:

Mode 1 contribution to $X_2 = \Psi_{21}\underline{X}_1$ (7-14)
Mode 2 contribution to $X_2 = \Psi_{22}\underline{X}_2$ (7-15)
Mode 3 contribution to $X_2 = \Psi_{23}\underline{X}_3$ (7-16)
Mode 4 contribution to $X_2 = \Psi_{24}\underline{X}_4$ (7-17)

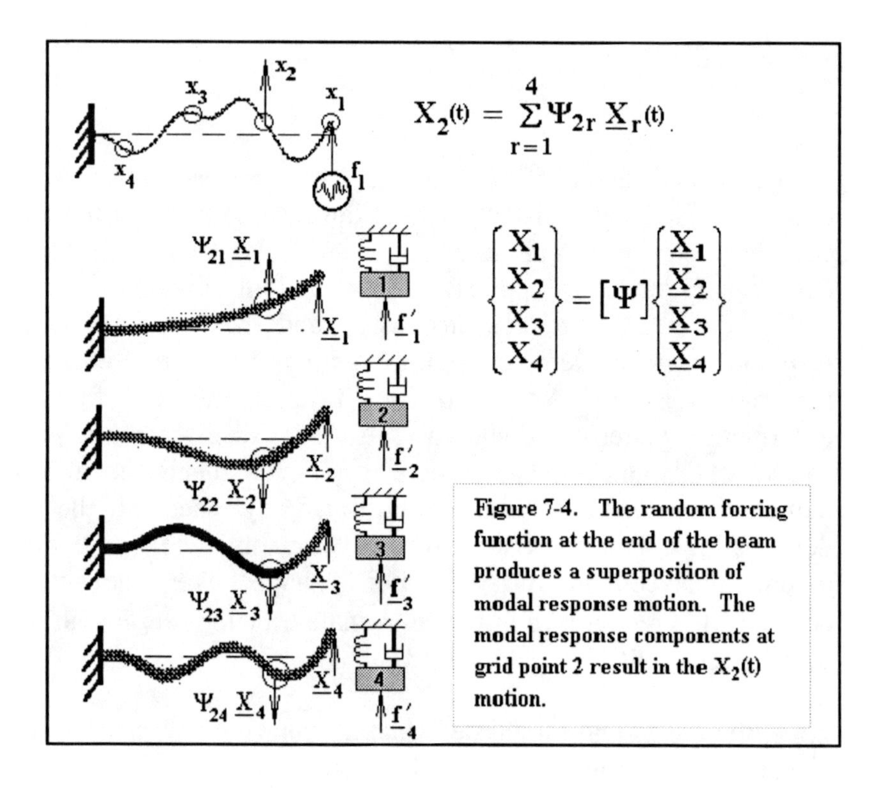


**Figure 7-4. The random forcing function at the end of the beam produces a superposition of modal response motion. The modal response components at grid point 2 result in the $X_2(t)$ motion.**

## 7.4 Superimposing Modal FRF's

We've seen now the picture of both physical displacements and physical forces as linear superpositions of modal displacements and contravariant modal forces. The concepts of the physical FRF and modal FRF have been discussed. This provides the physical intuition for the notion that the ratio of the physical response divided by physical force, i.e., the physical FRF, can also be seen as a superposition of modal FRF's. This is exactly the result already manifest as equation (6-20), repeated here:

$$[\, H(\omega) \,] = [\, \Psi \,][\, \underline{H}(\omega) \,][\, \Psi \,]^T \qquad (7\text{-}18)$$

where $\underline{H}(\omega)$ is diagonal.

Equation (7-18) will be expanded to obtain the formula for any one arbitrarily selected physical FRF, $h(\omega)_{jk}$. Expand (7-18) by selecting out the jth row of the premultiplying $[\,\Psi\,]$ matrix and the kth row of the $[\,\Psi\,]$ matrix to be transposed as $[\,\Psi\,]^T$ for post multiplying. The kth row will appear as a column after transposition.

$$\mathbf{h}_{\mathbf{jk}}(\omega) = \begin{pmatrix}\Psi_{j1} & \Psi_{j2} & \cdots & \Psi_{jn}\end{pmatrix} \begin{bmatrix} \ddots & & \\ & \underline{H} & \\ & & \ddots \end{bmatrix} \begin{pmatrix}\Psi_{k1} & \Psi_{k2} & \cdots & \Psi_{kn}\end{pmatrix}^T \tag{7-19}$$

Rewriting the transposed kth row as a column matrix, equation (7-19) becomes

$$\mathbf{h}_{\mathbf{jk}}(\omega) = \begin{pmatrix}\Psi_{j1} & \Psi_{j2} & \cdots & \Psi_{jn}\end{pmatrix} \begin{bmatrix} \ddots & & \\ & \underline{H} & \\ & & \ddots \end{bmatrix} \begin{Bmatrix}\Psi_{k1} \\ \Psi_{k2} \\ \vdots \\ \Psi_{kn}\end{Bmatrix} \tag{7-20}$$

First, multiply the right side column matrix by the diagonal $[\,\underline{H}\,]$ matrix.

$$\mathbf{h}_{\mathbf{jk}}(\omega) = \begin{pmatrix}\Psi_{j1} & \Psi_{j2} & \cdots & \Psi_{jn}\end{pmatrix} \begin{Bmatrix}\underline{h}_{11}\Psi_{k1} \\ \underline{h}_{22}\Psi_{k2} \\ \vdots \\ \underline{h}_{nn}\Psi_{kn}\end{Bmatrix} \tag{7-21}$$

Finally, multiply the jth row matrix times the r.h.s. column matrix of (7-21).

$$\mathbf{h}_{jk}(\omega) = \sum_{r=1}^{n} \Psi_{jr}\Psi_{kr}\underline{\mathbf{h}}_r(\omega) \tag{7-22}$$

It is quite clear that the physical FRF, corresponding to the physical displacement at D.O.F. j divided by physical force applied to D.O.F. k, is equivalent to the summation of modal FRF's, $\underline{h}_r$. The modal FRF's are summed up over all mode numbers, r=1,2,3,..., n, and each modal FRF is weighted by the product of the mode coefficients, $\Psi_{jr}$ and $\Psi_{kr}$.

And now comes the big payoff for all the earlier second order differential equation work in modal coordinates. Substituting the r.h.s. of equation (6-74) into $\underline{\mathbf{h}}_r(\omega)$ of (7-22),

$$\mathbf{h}_{jk}(\omega) = \sum_{r=1}^{n} \frac{\Psi_{jr}\Psi_{kr}}{\omega_r^{\,2}\underline{\mathbf{m}}_r} \frac{(1-\beta_r^{\,2}) - 2i\beta_r\zeta_r}{[(1-\beta_r^{\,2})^2 + 4\beta_r^{\,2}\zeta_r^{\,2}]} \tag{7-23}$$

We finally have the physically measurable FRF expressed in a formula that can actually be computed using parameters related to the modal properties of a structure. The physical FRF, $h_{jk}(\omega)$, can be computed for a dynamic structure provided that you can obtain resonance frequencies, modal damping factors, modal masses and all mode coefficients for each of the structural modes.

This is certainly one of the most important equations in experimental modal analysis and probably should be memorized by any test engineer or dynamist working in experimental structural dynamics on a day to day basis.

Figure 7-5 leverages the insights gained from equation (7-23) and the discussion around Figures 7-1 through 7-4. Figure 7-5 illustrates the use of modal FRF superposition to describe vibration displacements around a structure due to an excitation force at any single point of the structure.

Each mode is seen to have a modal representation as a simple SDOF (Single Degree Of Freedom) modal mass ($\underline{m}_r$), modal stiffness ($\underline{k}_r$) and modal damping ($\underline{c}_r$). Consequently, using the FRF modal coordinate representation, $\underline{h}_r(\omega) = \underline{x}_r(\omega)/\underline{f}'_r(\omega)$, each mode manifests the frequency response characteristics of the simple SDOF FRF as shown adjacent to each mode shape in Figure 7-5. The resonance frequencies of the four modes illustrated are indicated as $\omega_1$, $\omega_2$, $\omega_3$ and $\omega_4$.

The explicit graphical superposition of the four modal FRF's of Figure 7-5 to yield a physical FRF is shown in Figure 7-6. The four modal FRF's to be summed are the dashed and dotted curves of Figure 7-6, and the solid curve represents the result of the graphical summation. The solid curve is of course the physical FRF corresponding to displacement response at physical point number 2 on the example cantelever beam due to the force applied at physical point number 1.

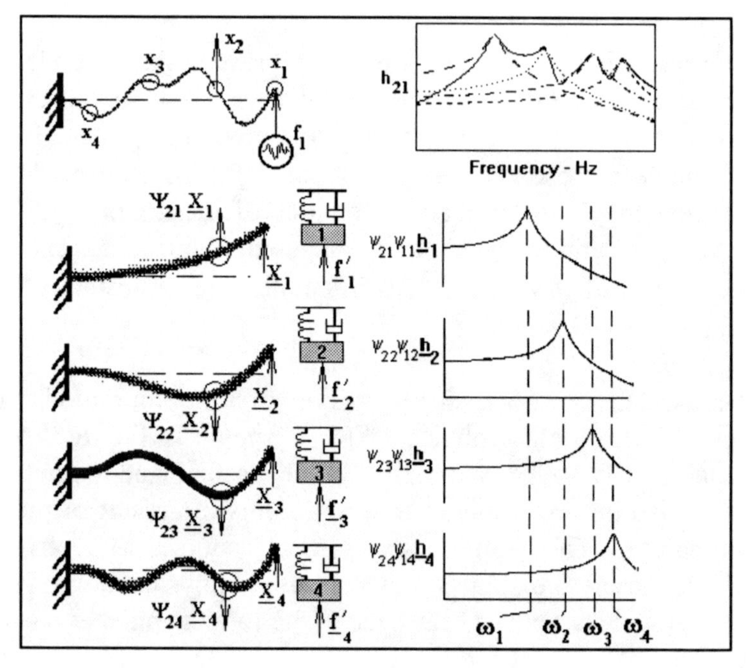


**Figure 7-5 Putting physical single point excitation and response in the context of modal FRF superposition. The physical force is seen as a superposition of modal forces, the physical displacement response is seen as a superposition of modal displacements, and the ratio of physical response to physical force, the FRF, is seen as the sum of modal FRF's weighted by the driving point and response point mode coefficients, $\Psi_{jr}$ and $\Psi_{kr}$.**

The peaks in the solid physical coordinate FRF are easily understood. They correspond to the resonance frequencies of each of the structure modes. But, notice the dip in the curve at approximately 6 Hz. This results from the effect of modal phase angles on the superposition.

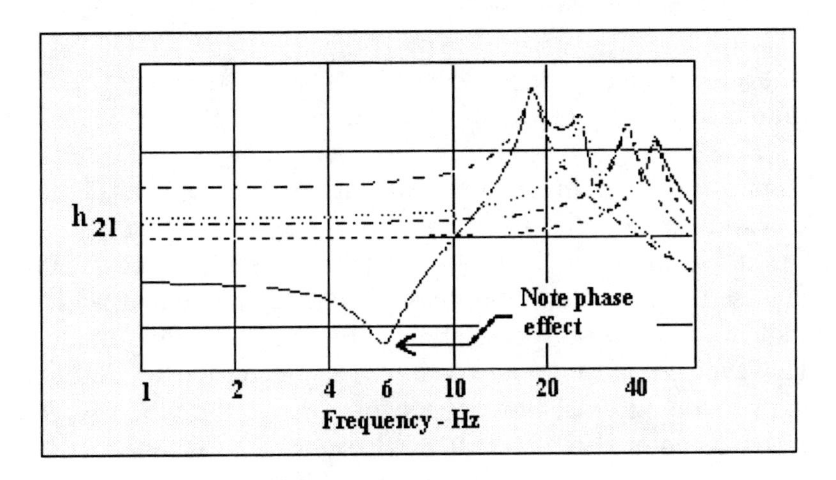


**Figure 7-6 Four SDOF Modal FRF's are summed to produce the physical FRF for displacement response point number two and excitation force at point number 1. A dip appears at approximately 6 Hz due to phase effects in the superposition.**

Recall that the modal FRF function of frequency is mathematically complex. The function has real and imaginary parts as plotted earlier in Figures 6-5 and 6-6. And the modal FRF phase function would be represented mathematically as

$$\underline{\theta}_r = \tan^{-1}(\underline{h}_{imag}/\underline{h}_{real}) \qquad (7\text{-}24)$$

The modal displacement response is predominantly in-phase with the modal force over the frequency range from zero Hz up to the neighborhood of the resonance frequency; and above the resonance frequency the modal displacement is predominantly out-of-phase with modal force. The phase angle ranges from zero to –90 degrees up to the resonance frequency, then ranges from –90 degrees to –180 degrees. The modal FRF superposition is actually implemented by summing the real parts, then summing the imaginary parts separately to obtain the physical coordinate FRF as a real and imaginary function.

The modal FRF superposition with phase effects can be visualized graphically using the modal FRF magnitude plots if we indicate the in-phase and out-of-phase frequency ranges and consider the summation with these conditions taken into effect. Figure 7-7 is constructed to assist in visualizing the graphical summation of the first three modal FRF's of our example cantelever beam. Plus signs are used to indicate the in-phase frequency range and negative signs indicate the out-of-phase range for each of the modal FRF's. However, it must be noticed that signs of the mode coefficients corresponding to displacements for each of the mode shapes at the response point also affect the in-phase or out-of-phase condition. This must be the case since the response mode coefficients, $\Psi_{jr}$, for each mode shape are present as a factor in the FRF superposition equation (7-23).

The FRF superposition equation appears again at the top of Figure 7-7

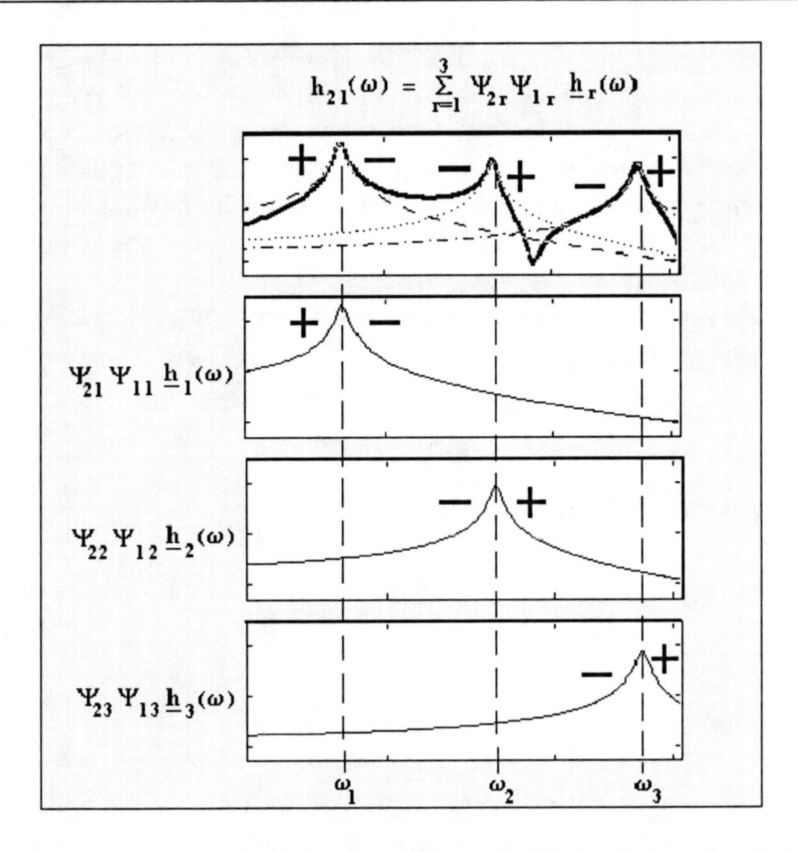


**Figure 7-7 Graphical summation of the first three modal FRF's of the cantelever beam example. The mode coefficients correspond to the physical FRF for force applied at point number one and response displacement at point number two. The in-phase and out-of-phase frequency ranges are indicated by the plus and minus signs. When curves have opposite signs over a given frequency range, the canceling effect pulls the resultant physical FRF down into a sharp dip as happens in the frequency range from $\omega_2$ to $\omega_3$. Over the range from $\omega_1$ to $\omega_2$, all three of the modal FRF's have an out-of-phase condition resulting in boosting up the value of the physical FRF over that range.**

The frequency domain is divided into ranges bounded by the resonance frequencies as illustrated in Figure 7-7. The modal FRF curves are graphically added together or subtracted within each of these frequency ranges, depending on the phase indicator plus or minus signs. The physical FRF is graphically developed for excitation force at the cantelever beam point number one and for displacement response at point number two.
This requires the use of mode coefficients, $\Psi_{1r}$ (excitation point) and $\Psi_{2r}$ (point two response).

## 7.4 Modal FRF Imaginary Part

Recall that the modal FRF is the sum of a real part and imaginary part:

$$\underline{h}_r(\omega) = \underline{h}_r(\omega)_R + i\underline{h}_r(\omega)_I \tag{7-25}$$

The equation for just the imaginary part of the modal FRF may be pulled out of equation (6-74):

$$\underline{h}_r(\omega)_I = \frac{-2\beta_r\zeta_r}{\omega_r^{\,2}\underline{m}_r\left[\left(1-\beta_r^{\,2}\right)^2 + 4\beta_r^{\,2}\zeta_r^{\,2}\right]} \tag{7-26}$$

This function was plotted in Figure 6-7 and indicates that the imaginary modal FRF peaks negatively at the resonance frequency (note the negative sign in equation 7-26).

In the same way we superimposed modal FRF magnitudes to yield a physical FRF (Figure 7-6), we can superimpose the modal FRF imaginary parts to get a physical FRF imaginary function. This process is illustrated in Figure 7-8. This example is again taken from our cantelever beam, representing the physical FRF, $h_{21}(\omega)$, obtained by applying an excitation force to the end of the beam (point one) while measuring response motion at point two.

The imaginary FRF is of particular interest to us because of the significance of the positive and negative peak values. These are the imaginary FRF values at the resonance frequencies. It will be seen that these values may be used to define mode coefficients and establish values for modal mass and modal stiffness. Therefore, we shall look closely at the expressions for values of the imaginary FRF at the resonance frequencies.

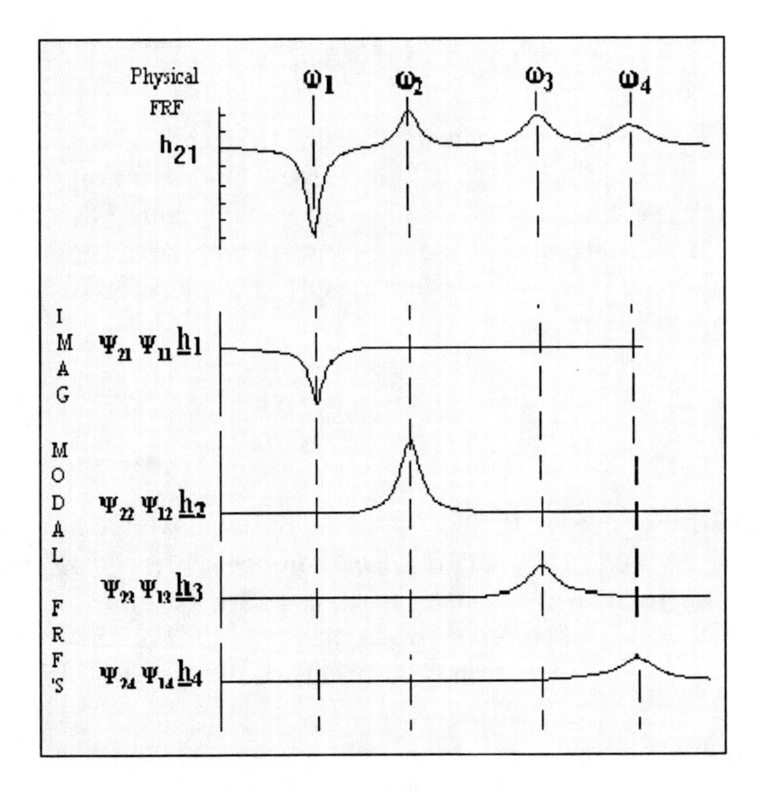


**Figure 7-8. Superposition of imaginary parts of modal FRFs. This example uses the physical FRF corresponding to response at cantelever beam point number two due to force applied at point number one.**

The expression for the value of $\underline{\mathbf{h}}_r(\omega_r)_I$, the peak modal FRF value at resonance, is obtained by setting $\omega = \omega_r$, or $\beta_r = 1$ in equation (7-26). Remember that $\beta_r = \omega/\omega_r$, the ratio of any frequency across the spectrum to the resonance frequency.

$$\underline{h}_r(\omega_r)_I = \frac{-2\zeta_r}{\omega_r^{\ 2}\underline{m}_r[4\zeta_r]} \tag{7-27}$$

or,

$$\underline{h}_r(\omega_r)_I = \frac{-1}{2\omega_r^{\ 2}\underline{m}_r\zeta_r} \tag{7-28}$$

Now, let's look at the value of the directly measured physical FRF at the resonance frequency of any given mode number r. For this example we choose what is called the driving point FRF. That is the FRF for which response motion is measured at the same point that the driving force is applied. Assuming some driving point, k, the physical driving point FRF is

$$h_{kk}(\omega) = \frac{x_k(\omega)}{f_k(\omega)} \tag{7-29}$$

When the resonance frequencies of a physical FRF are far apart, then the FRF value at any one resonance frequency, $\omega_r$, is determined to good approximation by just one modal FRF (with associated mode coefficients). All but one term in the modal FRF summation may be set to zero in such a case. The result for our physical driving point FRF is

$$h_{kk}(\omega_r) = 0 + 0 + 0 ... + \psi_{kr}\psi_{kr}\underline{h}_r(\omega_r)_I + 0 + 0 + 0 ... \tag{7-30}$$

or,

$$h_{kk}(\omega_r) = \psi_{kr}^{\ 2}\underline{h}_r(\omega_r)_I \tag{7-31}$$

This is simply the mathematical statement of the graphical picture of Figure 7-8, if you compare the modal FRF peaks at each resonance frequency ($\omega_1$, $\omega_2$, $\omega_3$, $\omega_4$) to the corresponding peaks in the physical FRF.

Substituting the r.h.s. of equation (7-28) in place of $\underline{h}_r(\omega_r)_I$ in equation (4-31),

$$h_{kk}(\omega_r) = -\psi_{kr}^{\ 2}\left[\frac{1}{2\omega_r^{\ 2}\underline{m}_r\zeta_r}\right] \qquad (7\text{-}32)$$

## 7.5 Scaling Modal Mass and Mode Coefficients

At this point the values of $\psi_{kr}$ and $\underline{m}_r$ are arbitrary. The resonance frequency, $\omega_r$, and damping ratio, $\zeta_r$, are fixed within the measured FRF, $h_{kk}(\omega_r)$. The frequencies are directly manifest, coinciding with the imaginary FRF positive or negative peaks. The damping ratio is computed using the resonance frequency along with structural bandwidth. The structural bandwidth, SB, is obtained from the real part of the FRF as illustrated in Figure 6-8. Repeating the damping ratio formula of equation (6-92),

$$\zeta_r = \frac{SB}{2\omega_r} \qquad (7\text{-}33)$$

We are free to fix either the reference mode coefficient, $\psi_{kr}$ or the modal mass, $\underline{m}_r$. If we decide to arbitrarily select a value for $\psi_{kr}$, then the value of $\underline{m}_r$ would be defined by equation (7-32):

$$\underline{m}_r = \frac{-\psi_{kr}^{\ 2}}{h_{kk}(\omega_r)}\left[\frac{1}{2\omega_r^{\ 2}\zeta_r}\right] \qquad (7\text{-}34)$$

Throughout this text we have been drawing pictures of mode shapes, sometimes arbitrarily specifying a mode coefficient value at some arbitrarily selected reference point. But, we were never forced to accept any specific value for the reference mode coefficient. However, once a value was established for a mode shape at any one selected Degree Of Freedom (physical point and direction on the structure), all remaining mode coefficients for that mode shape were fixed. The *shape* of the modal deformation is fixed, but that mode shape may be pictured with any arbitrarily selected deformation amplitude.

Analysts engaged in Finite Element Modeling often specify mode coefficient values equal to 1.0 at the maximum deformation Degree Of Freedom found on the structure for each mode shape. Another, perhaps more popular scaling for the analyst, is to set modal mass, $\underline{m}_r$, equal to 1.0 for each mode. The modal mass matrix, [ $\underline{M}$ ] is then the identity matrix, [ I ]. This fixes the values for $\psi_{kr}$ in equation (7-34).

The most natural scaling available to the experimentalist comes directly from the measured physical FRF itself. Simply set the mode coefficient values equal to the imaginary FRF peak values at the corresponding resonances:

$$\psi_{kr} = h_{kk}(\omega_r) \tag{7-35}$$

This defines the modal mass of equation (7-34) as

$$\underline{m}_r = \frac{|h_{kk}(\omega_r)|}{2\omega_r^2\zeta_r} \tag{7-36}$$

Or, replacing the $h_{kk}(\omega_r)$ in equation (7-36) with $\psi_{kr}$,

$$\underline{m}_r = \frac{|\psi_{kr}|}{2\omega_r^2\zeta_r} \tag{7-37}$$

Further, the modal stiffness, $\underline{k}_r$, is immediately available by solving for $\underline{k}_r$ in equation (7-34):

$$\underline{k}_r = \omega_r^{\,2}\,\underline{m}_r \tag{7-38}$$

Let's summarize where we are now. It appears that we have uncovered the wherewithal to compute any arbitrary physical FRF, $h_{jk}(\omega)$. We do it using equation (7-23) after obtaining values for the global modal parameters and the mode coefficients.

We refer to the parameters, resonance frequency, modal damping, modal mass and modal stiffness as global parameters. Those parameters show up, for the most part, anywhere on the structure. You should be able to measure an FRF between any two points on the structure and usually find all of the same resonance frequencies and modal damping values. Modal mass is global, although we rescale the values to suit the particular reference point selected for applying the excitation force (a consequence of having different reference point mode coefficient values). The individual mode coefficients are not global, being associated with specific points on the structure. The mode shape itself is typically global in character, however. You look across the length of the beam (the global structure) to see what a particular mode shape looks like.

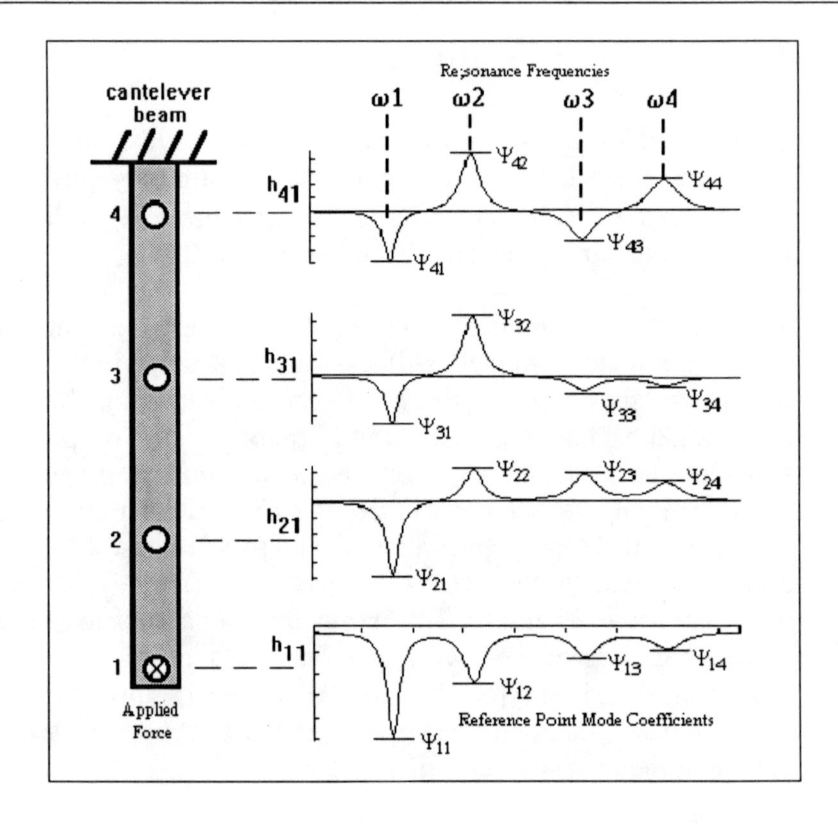


**Figure 7-9 Imaginary FRF's acquired from four points on the example cantelever beam. Resonance frequencies and all mode coefficients may be obtained directly from just one set of FRF's. Force is applied at a single reference point. $\Psi_{11}$, $\Psi_{12}$, $\Psi_{13}$ and $\Psi_{14}$ are the reference mode coefficients against which modal masses and all remaining mode coefficients are scaled.**

By taking mode coefficients from the peaks of a single set of FRF's (applying vibration excitation at just one point), and using equation (7-37) for computing modal mass, any of
the remaining FRFs of the full matrix of FRFs, [ H ], may be computed. Figure 7-9 schematically illustrates this ability to acquire all modal parameters and all mode coefficients from just one set of measured FRF's. FRF's are measured for each point on the beam with the excitation force applied at the end of the beam (point number one).

## 7.6 [ H ] Matrix and [ Ψ ] Matrix Applications

A brief preview of applications of the [ H ] and [ Ψ ] matrices is inserted here to emphasize their importance to structural dynamics. We wish to motivate later pursuit of more sophisticated methods of extracting mode shapes from experimental FRF data. It will be seen that structures are generally quite a bit more complicated than our simple cantelever beam example. And complicated structures often yield FRFs with resonance frequencies too closely spaced for accurately obtaining mode coefficients directly from physical FRF imaginary peaks.

This ability to compute all FRFs of the full [ H ] matrix has significant implications. Given the full matrix of FRFs, it is possible to compute the response motion anywhere on a structure due to any arbitrarily specified set of applied dynamic forces:

$$\{\mathbf{X}(\omega)\} = [\ \mathbf{H}\ (\omega)\ ]\{\mathbf{F}(\omega)\} \qquad (7\text{-}39)$$

The original set of forces to be applied to the structure may be defined either in the time domain or the frequency domain. If the forces are given as functions of time, you simply Fourier Transform the forces for entry into the force column matrix of equation (7-39).

If the forces are specified as spectra in the form of Fourier Transforms, then they are immediately ready for use in (7-39). Sometimes the forces are available only as Power Spectra or Power Spectral Density (PSD) spectra. That means that the phase vs. frequency has not been specified. There are a number of ways to specify a phase function and add it to an amplitude spectrum, eventually converting the spectra to the Fourier Transform data format. For example, converting a PSD spectrum to a Fourier Transform representing a steady state random force is straightforward.

In some cases it is important to know what dynamic forces are being applied to a structure during its normal operation. Perhaps a dynamics analyst is evaluating alternative product designs using Finite Element Models. It is often impossible to directly measure the forces.

However, it is usually fairly easy to measure response motion. After developing the [ H ] matrix, the matrix inverse is computed, and equation (7-39) is solved for the set of physical dynamic forcing functions.

$$\{F(\omega)\} = [H(\omega)]^{-1}\{X(\omega)\} \tag{7-40}$$

Sometimes it is known that forces are applied at a certain limited number of points on the structure. In such a case it would not be necessary to develop the full [ H ] matrix. Then, a pseudo inverse would be performed on the [ H ] matrix, and the number of degrees of freedom entering into the matrix equation could be considerably reduced.

Dynamics analysts quite commonly request the experimentalist to provide measured FRF's for direct comparison to the corresponding FRF's computed from the FEM (Finite Element Model). This allows the analyst to adjust the FEM to match the actual structure.

A very popular way of correlating the FEM with the test model is to perform what is called an orthogonality check:

$$[\Psi]^{T}_{Test}[M]_{FEM}[\Psi]_{FEM} \overset{?}{=} \begin{bmatrix} \ddots & & \\ & \underline{M} & \\ & & \ddots \end{bmatrix} \tag{7-41}$$

If the FEM mode shapes match the test mode shapes, the r.h.s. of equation (7-41) will be the diagonal modal mass matrix. A perfect match is never achieved, and sometimes the correlation is considered satisfactory if the off-diagonal elements of the r.h.s. have values equal to 0.1 or less (assuming the modal masses on the diagonal have been scaled to a value of 1.0 (implying a rescaling of the test mode coefficients in favor of FEM practice).

# CHAPTER VIII

# MODAL ANALYSIS OVERVIEW

## 8.1 Experimental Modeling

Structural dynamics technology includes such activities as Finite Element Modeling, Experimental Modeling, Loads and Response Analysis, Operation Testing and Environmental Shock and Vibration Testing. Here, of course, we are focused on Experimental Modeling. Modal analysis, the key component of experimental modeling, has become an important tool in the systematic approach to designing products for successful performance in an environment of mechanical shock and vibration. Indeed, the growing applications of experimental structural dynamics have led to the appearance of modal test laboratories throughout the domestic appliance, industrial, automotive and aerospace industries.

A modal test model consists of an adequate set of global modal parameters and all associated mode shapes (the [ $\Psi$ ] matrix). A structure is completely defined (dynamically) by this model. Actually, the completeness may be limited by the number of modes, frequency range and number of structure points included in the model.

A full [ H ] matrix also, in principle, completely describes a structure and could be regarded as a model. It should be noted that it is virtually impossible to produce a model by measuring all FRFs of the [ H ] matrix. An electrodynamic shaker is typically used to apply the excitation force. This piece of equipment might occupy a cylindrical volume of eight or ten inches in length and five or six inches in diameter. And it must be supported from a mounting point that is completely decoupled from the test structure. For many structures, the majority of points would be inaccessible to a shaker.

Remember, each shaker location provides FRF measurements making up one column of the [ H ] matrix. You would have to move the shaker to every point and direction on the structure to fill out the [ H ] matrix. Aside from the inaccessibility, the shear number of measurements makes filling out the [ H ] matrix impractical. A modal test on the NASA Orbiter OMS (Orbiter Maneuvering System) pod included approximately 2400 FRFs for just one column of the [ H ] matrix and required a three to four week effort. This circumstance leads to one of the major uses of the modal test model, that is to generate a more complete [ H ] matrix using equation (7-23).

## 8.2 The Direct Peak Extraction Method

We began by developing the understanding of summing modal FRFs to obtain a physically measurable FRF. The modal analysis process is just the reverse of this, decomposing an experimentally acquired FRF into separate modal FRFs.

By now, it is understood that there are two broad steps in the experimental modal analysis process: 1) Acquire at least one set of FRFs (one column of the [ H ] matrix) and 2) Process the FRF data to obtain a modal test model.

So far, we have seen from our example cantelever beam that the resonance frequencies and mode shapes can be obtained directly from the imaginary parts of the FRFs. The mode shapes are immediately defined by the imaginary peaks as pictured in Figure 8-1 below. Mode shape number one, for example, is displayed as a curve drawn through the peaks of FRFs along the beam corresponding to the first resonance frequency, $\omega_1$.

Keep in mind that all of the FRFs used here have been obtained from applying a vibratory excitation force at just one point on the structure (the end of the cantelever beam). By measuring response motion over the entire structure it was possible to define all mode shapes. The relationships among the physical locations on the structure, the [ H ] matrix and the [ Ψ ] matrix are further clarified in Figure 8-2.

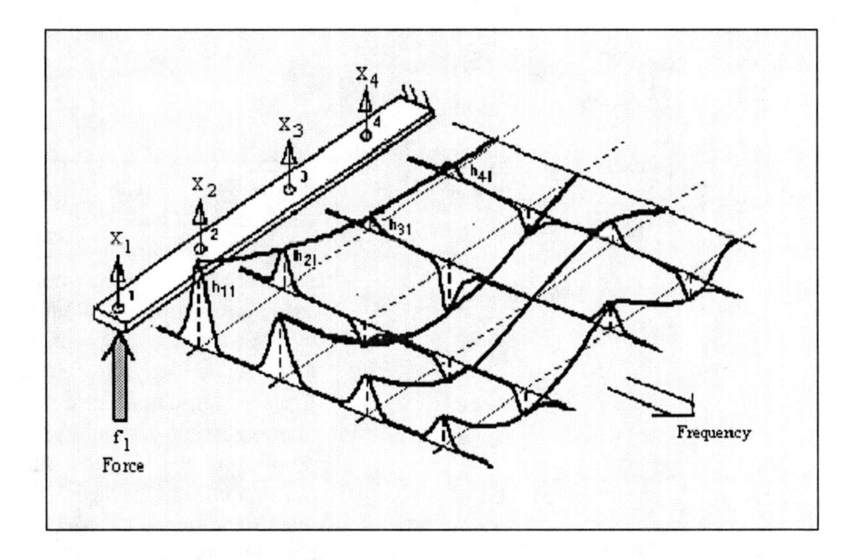


**Figure 8-1 The connection between mode shapes and imaginary FRF positive and negative peaks are pictured. FRF's are plotted vs. frequency for each of the four points on the cantelever beam. A mode shape deformation pattern appears when tracing out a curve through all FRF peaks corresponding to a selected resonance frequency.**

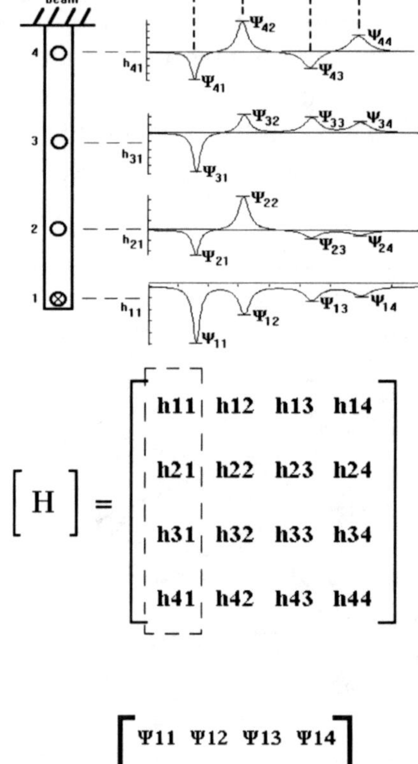


$$
\left[\Psi\right] = \begin{bmatrix} \Psi 11 & \Psi 12 & \Psi 13 & \Psi 14 \\ \Psi 21 & \Psi 22 & \Psi 23 & \Psi 24 \\ \Psi 31 & \Psi 32 & \Psi 33 & \Psi 34 \\ \Psi 41 & \Psi 42 & \Psi 43 & \Psi 44 \end{bmatrix}
$$

Mode Shape Numbers
4
3
2
1

Figure 8-2 Cantelever beam configuration with four measurement points. Imaginary parts of FRF's are shown in association with measurement points and a spectral matrix of FRF's. The FRFs shown result from response at all grid points due to a single excitation force at point number 1. This produces the first column of the FRF matrix (dashed box). Mode coefficients correspond to peaks at the resonance frequencies (a good approximation when modes are well separated in frequency.

Experimental modal analysis technology provides a number of different methods for extracting global modal parameters and mode shapes from measured FRFs. We shall refer to the method described by Figure 8-2 as the *direct peak extraction* method. This is the easiest and quickest method of obtaining modal parameters and mode shapes using a computer program, once the FRFs have been acquired. The software utilizes a simple peak search algorithm. The peak search algorithm identifies resonance frequencies and mode coefficients.

Just one FRF, the driving point FRF, would be required to obtain the global modal parameters. The peak-valley search algorithm is applied to the real part of the driving point FRF to obtain the modal damping factor, $\zeta_r$, for each of the modes (equation 7-33).

## 8.3 Problems With The Peak Extraction Method

But there is a serious problem with using the direct peak extraction method as a general tool for developing a modal test model from measured FRFs. Figure 8-3 displays a measured imaginary FRF (solid curve) exhibiting three resonance frequencies. The first resonance peak is negative, followed by a second positive resonance peak whose frequency is very close to the first. The three pure modal FRFs (scaled by the proper mode coefficients, i.e., $\psi_{jr}$ $\psi_{kr}$ $\underline{h}_r$) are overlaid for comparison to the measured FRF. Notice that the peaks of the first two modal FRFs are not at all coincident with the corresponding measured FRF peaks. The third resonance frequency is well separated from the first two resonances, and the measured FRF peak very accurately matches the modal FRF peak in this case. Clearly, the direct peak extraction method applied to the first two resonance frequencies on the measured FRF would not yield correct values for the mode coefficients.

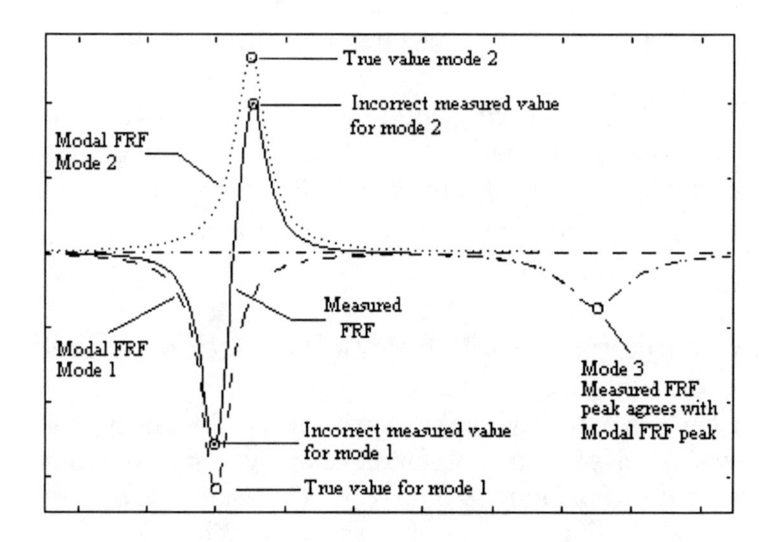


**Figure 8-3 A measured FRF is plotted with overlays of modal FRFs. The first two resonance frequencies are close together, and the measured imaginary peaks do not accurately represent the mode coefficients. This motivates other methods for extracting mode coefficients from measured FRFs.**

The reason for the mismatch between the measured imaginary FRF peaks and the modal FRF peaks is illustrated in Figure 8-4. The measured negative peak at the first resonance would match the first modal FRF peak, except for the effect of the mode two FRF. And the measured positive peak at the second resonance would match the second modal FRF peak, except for the effect of the mode one FRF. Remember that all of the modal FRFs must be summed together to produce the measured FRF (equation 7-23). Equation (7-30) works only for modes that are widely separated in frequency.

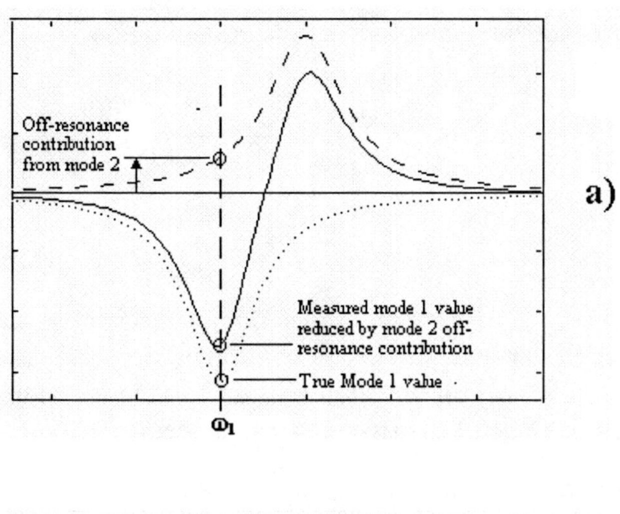


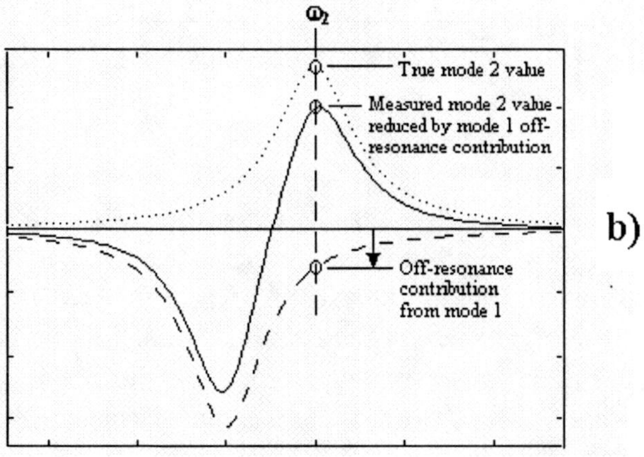


**Figure 8-4 The reason for the mismatch between peak values of a physical FRF and modal FRF peak values. The upper overlaid plots, a), show the way in which the off-resonance contribution from mode 2 adds a positive increment to the otherwise negative mode 1 FRF value. Likewise, referring to b), the negative off-resonance contribution of mode 1 results in a reduced physical FRF value for the mode 2 peak.**

Consider the imaginary part of the equation (7-23) summation:

$$h_{jk}(\omega)_I = \sum_{r=1}^{n} \frac{\psi_{jr}\psi_{kr}}{\omega_r^2 \underline{m}_r} \left[ \frac{-2\beta_r\zeta_r}{(1-\beta_r^2)^2 + 4\beta_r^2\zeta_r^2} \right] \qquad (8\text{-}1)$$

Evaluate the physical FRF, equation (8-1), at the resonance frequency, $\omega_1$, of the first mode. Compute this physical FRF value by summing over just the first three modes, r = 1,2,3 (assume all higher modes have negligible contribution to the measured FRF in the frequency range of the first two resonances).

$$h_{jk}(\omega_1)_I = -\frac{\psi_{j1}\psi_{k1}}{2\omega_1^2 \underline{m}_1\zeta_1} + \frac{\psi_{j2}\psi_{k2}}{\omega_2^2 \underline{m}_2} \left[ \frac{-2\beta_2\zeta_2}{(1-\beta_2^2)^2 + 4\beta_2^2\zeta_2^2} \right] + 0$$

(8-2)

where $\beta_2 = \omega_1/\omega_2$. Also, the off-resonance response of mode number three, along with all higher frequency modes, makes a negligible contribution to the total measured FRF at the resonance frequency of the first mode. However, the second modal FRF makes a significant contribution. It is seen that as long as the ratio, $\omega_1/\omega_2$, is in the neighborhood of 1.0, then the second term in (8-2) may not be ignored. But, when the $(1-\beta_2^2)^2$ expression is large compared to the $2\beta_2\zeta_2$ expression, then the second mode becomes insignificant.

Similarly, the measured FRF at the second resonance frequency is

$$h_{jk}(\omega_2)_I = \frac{\psi_{j1}\psi_{k1}}{\omega_1^2 \underline{m}_1} \left[ \frac{-2\beta_1\zeta_1}{(1-\beta_1^2)^2 + 4\beta_1^2\zeta_1^2} \right] - \frac{\psi_{j2}\psi_{k2}}{2\omega_2^2 \underline{m}_2\zeta_2}$$

(8-3)

where $\beta_1 = \omega_2/\omega_1$, and this ratio is getting close to a value of one, since the two resonance frequencies are close to each other. It is said, in reference to equation (8-3), that the off-resonance response of mode number one makes a contribution to the measured FRF at the mode two resonance frequency. Since $\psi_{j1}$ and $\psi_{j2}$ have opposite signs, the measured FRF peak at the second resonance frequency has been reduced. This circumstance of closely spaced modes (near-neighbor resonance frequencies) defeats the ability to obtain true mode coefficient values directly from the imaginary peaks. It must be said that most real structures are afflicted by this condition.Thus, the direct peak extraction method is often invalid as a strategy for modal analysis.

## 8.4 Modal Analysis Background

Structural dynamics theoreticians have looked for mathematical methods of extracting modal parameters and mode coefficients from measured FRFs since the days of World War II aircraft research. A landmark paper was authored by Kennedy and Pancu in 1947. In the early 1970's Albert Klosterman of Structural Dynamics Research Corporation (SDRC) and Mark Richardson of Hewlett Packard (presently President, Vibrant Technology, Inc.) developed competing commercial modal analysis software for the working test engineers. This followed Klosterman's much referenced 1971 doctoral dissertation, "On The Experimental Determination And Use Of Modal Representations Of Dynamic Characteristics" at the University Of Cincinnati.

During that same time period at the University Of Cincinnati, Dr. Dave Brown began developing what has become a leading academic center of experimental structural dynamics. After an encounter with Spitznogle at a technical conference, Dr. Brown played a key role in introducing the Spitznogle complex exponential curve-fitting algorithm as a modal analysis tool. Klosterman commented on the algorithm in his thesis and subsequently included a fully developed version in the SDRC modal analysis software. Spitznogle, a researcher at Texas Instruments, had developed the method for analyzing submarine sonar signals under contract to the U.S. Navy.

Modal analysis software allows the engineer to work at a computer interactively with on-screen graphics and a modern windows-based user interface to develop modal test models from FRF data. Once mode shapes have been obtained, they may be visualized with animated graphics displays (a novel feature of both Klosterman's and Richardson's original software). Havard Vold of SDRC developed more sophisticated curve-fitting methods currently used to extract the pure modal values in the presence of distorted data and closely spaced frequencies. Efforts to enhance these methods as well as develop new strategies continue to this day. A major forum in the current field of analytical and experimental structural dynamics is the International Modal Analysis Conference (IMAC), sponsored jointly by Union College of Schenectady, NY and Society For Experimental Mechanics, Inc., Bethel, Connecticut.

## 8.5 Curve Fitting

Before taking up specific modal analysis examples, we briefly review the general concept of curve fitting.

By curve fitting we refer to the mathematical process of adjusting the parameters in an equation so that the graph of the equation matches a particular set of data. As an example, consider the equation of a straight line. The point-slope form of the straight-line equation is

$$Y = mX + b \tag{8-4}$$

A graph of equation (8-4) is sketched in Figure 8-5. There are of course an infinite number of straight lines that could be plotted from equation (8-4), depending on the values selected for the two parameters, m and b. The parameter, m, defines the slope of the line and the parameter, b, establishes the Y-axis intercept where X = 0.

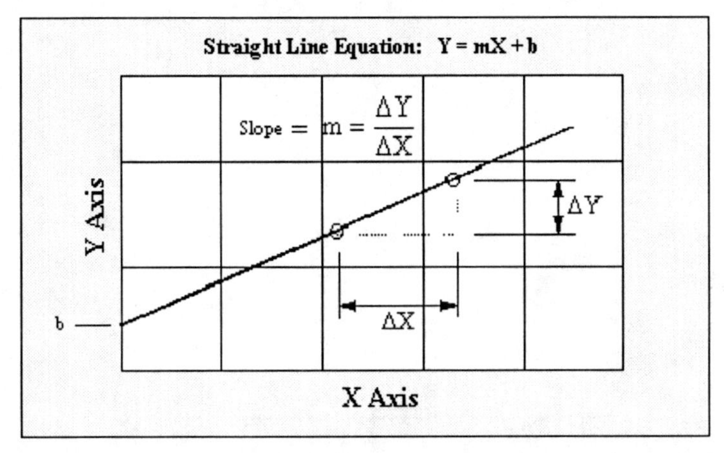


**Figure 8-5 A straight line is defined by an equation having two parameters, m for defining the slope of the line and b for defining the intercept point on the Y axis.**

Now, consider the simplest example of curve fitting. An experimentalist has two measurement points making up his set of data. Point 1 has values, $X_1$ and $Y_1$ and data point 2 has values $X_2$ and $Y_2$. This situation is posed in Figure 8-6 where we have two data points, upper sketch a), and an equation is found that provides the straight line that fits the two points exactly, lower sketch b).

The straight-line equation is found by solving two simultaneous equations. The two simultaneous equations may be written as follows, one equation for each of the known X and Y values for points one and two. Substituting the point values into equation (8-4),

$$Y_1 = mX_1 + b \tag{8-5}$$

$$Y_2 = mX_2 + b \tag{8-6}$$

These two equations may be written in matrix form as follows:

$$\begin{Bmatrix} Y_1 \\ Y_2 \end{Bmatrix} = \begin{bmatrix} X_1 & 1 \\ X_2 & 1 \end{bmatrix} \begin{Bmatrix} m \\ b \end{Bmatrix} \qquad (8\text{-}7)$$

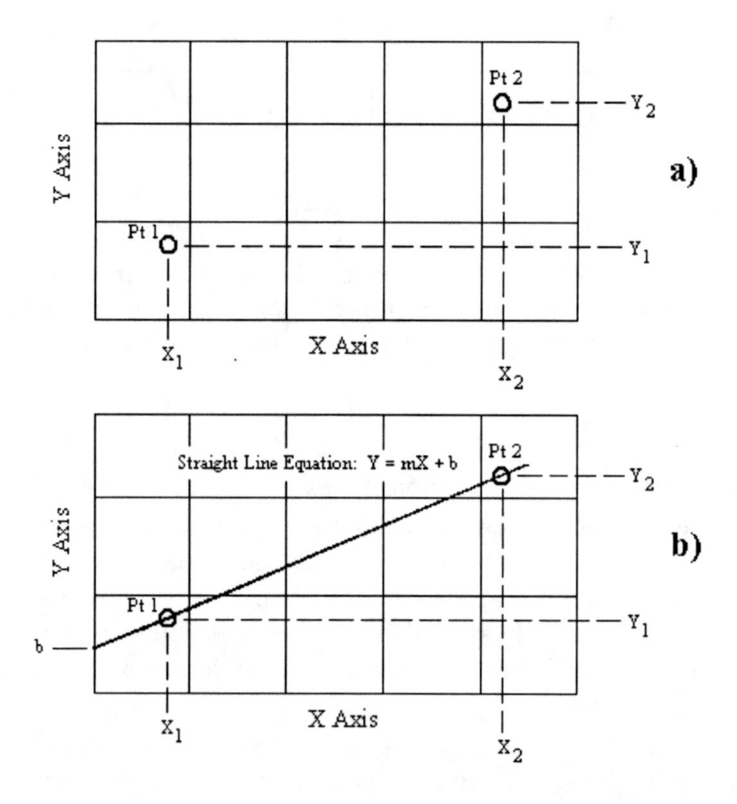


**Figure 8-6 The simplest example of curve fitting. Two data points are fit exactly with a straight-line equation.**

Equation (8-7) may be solved for the parameters m and b by multiplying both sides of the equation by the inverse of the square matrix, resulting in

$$\begin{Bmatrix} m \\ b \end{Bmatrix} = \begin{bmatrix} X_1 & 1 \\ X_2 & 1 \end{bmatrix}^{-1} \begin{Bmatrix} Y_1 \\ Y_2 \end{Bmatrix} \tag{8-8}$$

There are many different types of equations for the varied types of curves which have exact solutions, often expressed in matrix form similar to equation (8-8). Differing numbers of parameters and numbers of equations would be required, depending on the type of curve. The equation of a circle could be matched to three points (assuming they were not collinear). A polynomial could be used to match data that seems to swing up and down through a sequence of peaks and valleys.

But what if we have more data points in our set of data than is required to fit a particular equation? Returning to the straight-line equation example, what if we have a data set consisting of five points, and we know that a straight-line equation is the correct way to represent the data analytically? Measured data always includes some component of error. The true analytical representation of the data must somehow average its way through the data. We need to make use of all of the data points, even though equation (5-8) can only accommodate two data points.

Figure 8-7 depicts the straight-line curve fit of five data points. Here is the approach. Just write down the five simultaneous equations in matrix form, one equation for each data point, just like equation (8-7).

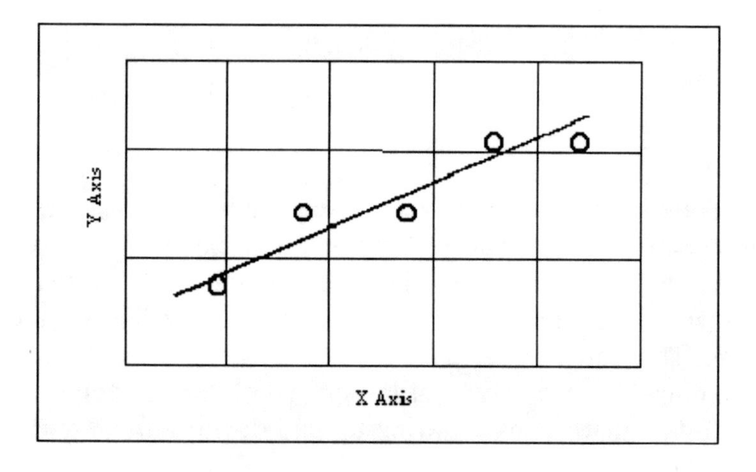


**Figure 8-7 A straight-line equation describes the curve averaging its way through five points. An overdetermined system of five straight-line equations were used to obtain the best fit.**

The equations in matrix form are:

$$\begin{Bmatrix} Y_1 \\ Y_2 \\ Y_3 \\ Y_4 \\ Y_5 \end{Bmatrix} = \begin{bmatrix} X_1 & 1 \\ X_2 & 1 \\ X_3 & 1 \\ X_4 & 1 \\ X_5 & 1 \end{bmatrix} \begin{Bmatrix} m \\ b \end{Bmatrix} \tag{8-9}$$

The slope, m, and the intercept, b, are solved for as before. Multiply both sides of equation (8-9) by the inverse of the rectangular matrix:

$$\begin{Bmatrix} m \\ b \end{Bmatrix} = \begin{bmatrix} X_1 & 1 \\ X_2 & 1 \\ X_3 & 1 \\ X_4 & 1 \\ X_5 & 1 \end{bmatrix}^{-I} \begin{Bmatrix} Y_1 \\ Y_2 \\ Y_3 \\ Y_4 \\ Y_5 \end{Bmatrix} \qquad (8\text{-}10)$$

Right away we have a special problem arising as to how to invert the matrix that is now rectangular, rather than square. A special inverse procedure known as the pseudo-inverse, or generalized inverse , will be used. This inverse is indicated in equation (8-10) by using the notation, -I, in the matrix exponent rather than the notation, -1. The development of the generalized inverse computation is as follows. Start again with equation (8-9) using a little more compact notation.

$$\{Y\} = [X \;\vdots\; 1]\{p\} \qquad (8\text{-}10)$$

The strategy will be to get the matrix equation in a form that leads to the inversion of a square matrix rather than one that is rectangular. A fruitful path to follow would be to multiply both sides of (8-10) by transpose of [ X | 1 ]:

$$[X \;\vdots\; 1]^T\{Y\} = [X \;\vdots\; 1]^T[X \;\vdots\; 1]\{p\} \qquad (8\text{-}11)$$

Realizing that any rectangular matrix multiplied by the same matrix transposed produces a square matrix, it is seen that equation (8-11) can now be solved for { p }.

$$\{p\} = \left[[X \;\vdots\; 1]^T [X \;\vdots\; 1]\right]^{-1} [X \;\vdots\; 1]^T \{Y\} \qquad (8\text{-}12)$$

The straight line defined by the m and b values coming out of this solution fits the data points as illustrated in Figure 8-7. This curve fit turns out to be the best fit in a least square sense.

Before leaving the discussion of equation (8-12) we will step through a sequence of dummy matrix operations to verify the consistency in the number of rows and columns within each matrix at each stage of processing. The rows and columns of equation (8-12) are represented as

$$\begin{Bmatrix} m \\ b \end{Bmatrix} = \left[ \begin{bmatrix} - & - \\ - & - \\ - & - \\ - & - \\ - & - \end{bmatrix}^T \begin{bmatrix} - & - \\ - & - \\ - & - \\ - & - \\ - & - \end{bmatrix} \right]^{-1} \begin{bmatrix} - & - \\ - & - \\ - & - \\ - & - \\ - & - \end{bmatrix}^T \begin{Bmatrix} - \\ - \\ - \\ - \\ - \end{Bmatrix} \qquad (8\text{-}13)$$

Carry out the dummy matrix product $[\,-\,]^T[\,-\,]$

$$\begin{Bmatrix} m \\ b \end{Bmatrix} = \begin{bmatrix} - & - \\ - & - \end{bmatrix}^{-1} \begin{bmatrix} - & - \\ - & - \\ - & - \\ - & - \\ - & - \end{bmatrix}^T \begin{Bmatrix} - \\ - \\ - \\ - \\ - \end{Bmatrix} \qquad (8\text{-}14)$$

Next, multiplying out the transposed rectangular times the column matrix,

$$\begin{Bmatrix} m \\ b \end{Bmatrix} = \begin{bmatrix} - & - \\ - & - \end{bmatrix}^{-1} \begin{Bmatrix} - \\ - \end{Bmatrix} \tag{8-15}$$

Finally, it is clear that the matrix multiplication on the right yields a column matrix with two rows.

$$\begin{Bmatrix} m \\ b \end{Bmatrix} = \begin{Bmatrix} - \\ - \end{Bmatrix} \tag{8-16}$$

The simple straight-line curve fit required solutions for just two simple parameters, m and b. Our actual measured FRF data will be fitted with analytical FRFs. Those functions will be found by solving for the global modal parameters, $\omega_r$, $\zeta_r$ and $\underline{m}_r$ along with the mode coefficients, $\psi_{kr}$ and $\psi_{jr}$. An example of FRF curve fitting results is shown in Figure 8-8. The test FRF data is displayed in scatter plot style (individual data points) and the analytical FRF's resulting from curve fitting are displayed as solid curves. The upper chart shows results of curve fitting each of four modes separately. The lower chart shows the final analytical FRF (solid curve) plotted over the test data for comparison. The final analytical FRF was obtained by summing the four modal FRF's (described by equation 7-23).

There are a number of different curve-fit methods and procedures available to the modal analysis process. The next section will organize some of these.

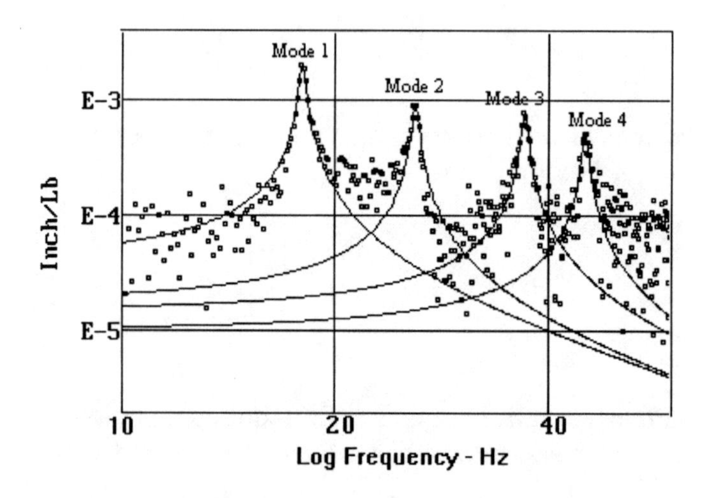


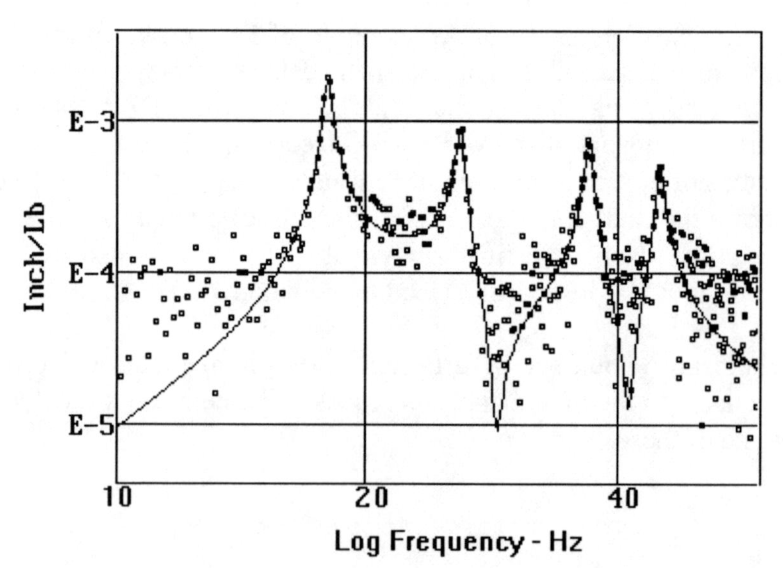


**Figure 8-8 Curve fitting measured FRF data (shown as scatter plots). The upper chart shows the results of first curve-fitting four modes individually. The lower chart shows result of the analytical FRF curve fit after modal FRFs have been superimposed. The log magnitude form of the FRF is used.**

## 8.6 Classification Of Mode Extraction Methods

It will not be the goal of this text to present an exhaustive account of all presently known mode extraction methods. The more common methods among those that are based on processing FRF data will be listed. Notable among methods not discussed in this text are the ARMA (Auto Regressive Moving Average), ERA (Eigensystem Realization Algorithm) and the ITD (Ibrahim Time Domain) methods. Table 8-1 outlines some of the main categories of the FRF mode extraction methods discussed here.

**Table 8-1. A classification of some of the FRF mode extraction methods employed in experimental structural dynamics.**

| **MODAL ANALYSIS METHODS** | | | |
|---|---|---|---|
| **Direct Extraction** | **Curve-Fitting** | | |
| | **Single Reference** | | **Multiple FRFs** |
| | **SDOF** | **MDOF** | |
| Peak Extraction | Circle Fit | Complex Exponential | Complex Exponential Polyreference |
| Complex Peak Extraction | Least Squares Circle Fit | Least Squares Complex Exponential | Orthogonal Polyreference |
| | Direct Parameter Polynomial Fit | Laplace Domain Partial Fraction | Multiple Response Direct Parameter |

The mode extraction methods are first divided into two groups: **Peak extraction** methods and **curve-fit methods**. The curve-fit methods are further divided into **single reference** and **multiple reference**. The single reference methods consist of those methods that utilize FRFs from just one column of the [ H ] matrix, acquired by measuring response motion at all points and directions on the structure while applying a vibratory force at a single point and direction. The multiple response, multiple reference methods perform simultaneous curve-fitting on multiple columns of the [ H ] matrix. That is, a complete set of FRFs are obtained for two or more shaker locations. The use of four shaker locations would not be uncommon when applying one of the multiple reference methods. This would result in a measured [ H ] matrix having four columns. Given about 100 response measurements, the [ H ] matrix would be rectangular, 100 x 4 (100 rows and 4 columns). The way in which all rows and columns may be simultaneously included in a curve-fit process will be discussed in a subsequent chapter of this text.

A drawback of the single reference method as compared to the multiple reference method involves a couple of factors. First, it is possible that applying a force at a single point will fail to adequately excite a few of the modes. Notice that equation (7-23) implies that if the applied force location is close to a node line (a spatial zero cross-over point) for a particular mode shape, then that mode will not be prominently excited. That is because the driving point mode coefficient, $\psi_{kr}$, would be close to zero, making the FRF value in equation (7-23) close to zero for that particular mode. Some of the resonance frequencies may not show up at all for a given excitation driving point.

In the case of the cantelever beam it was easy to see that all modes would be excited by a force applied at the end of the beam. All of the mode coefficients, $\psi_{kr}$, at the end of the cantelever beam were healthy nonzero numbers. But, for real-life structures having a large number of modes to be analyzed, it is difficult to find one point on the structure that has large values of $\psi_{kr}$ (a good shaker location) for every mode. Applying forces at four different locations offers a fair probability of exciting all modes of interest.

A second problem with curve-fitting FRFs acquired from a single reference point is that most structures are nonlinear to some extent. The resonance frequency and damping of a given mode are a little different, depending on the level of vibration displacement. Again, differing driving point mode coefficient values are encountered as you move a shaker around the structure to different drive points. And this means that the level of excitation for a given mode varies, depending on the drive point location. Thus, slightly different measurements of resonance frequency and modal damping will be observed for different shaker locations. Engineers usually like to obtain resonance frequency and modal damping values that are truly global, not dependent on the location of the applied force. A resonance frequency value for a given mode should be the one-and-only true value. Performing a single curve-fit process on FRFs from multiple shaker locations and multiple responses simultaneously yields such a global result.

Having made the case for global parameters, it should be said that there are instances when FRFs from a single reference point are desireable. For example, when evaluating the vibration effects of on-orbit thrust transients on payloads installed in the NASA Orbiter Payload Bay, it is useful to establish the FRFs between the point of an individual rocket's application of force and the relevant response points in the Payload Bay.

The single reference methods include two categories of curve-fit methods: The SDOF (Single-Degree-Of-Freedom) methods and the MDOF (Multiple-Degree-Of-Freedom) methods. Here, SDOF and MDOF refer specifically to modal Degrees-Of-Freedom (as opposed to points and directions on the structure).

## SDOF Curve Fitting

The upper chart of Figure 8-8 is an example of SDOF curve fitting. Each mode is fit separately, solving for the modal parameters of a single modal mass, spring and dashpot. The curve fit of each mode actually uses data points confined to the neighborhood of each resonance frequency. The data points are sometimes limited to those within the structural bandwidth of each mode. The detailed development of a single reference SDOF curve fit method, the Circle Fit, will be presented in Chapter 9.

## MDOF Curve Fitting

The MDOF curve fit utilizes data across a broad frequency range, encompassing two or more resonance frequencies. Incorporating data spanning the four resonance frequencies of Figure 8-8 would result in the direct simultaneous solution of all four sets of modal parameters. The curve fit result of the lower chart of Figure 8-8 would be obtained immediately, without the need to solve for the modes separately.

Most single reference MDOF methods perform the curve fit on one FRF at a time. The Multiple Response Direct Parameter Estimation method uses multiple FRFs simultaneously to solve for sets of global modal parameters.

## Multiple Response/Multiple Reference MDOF

Table 8-1 lists two Multiple Response/Multiple Reference MDOF methods, Complex Exponential Polyreference and Orthogonal Polyreference. Both of these methods utilize groups of FRFs from two or more columns of the test [ H ] matrix in a simultaneous curve fit of all selected data. Both methods perform simultaneous solutions for the global modal parameters ($\omega_r$, $\zeta_r$ and $\underline{m}_r$). Utilizing reference and response data from all over the structure in this manner provides a truly spatial global solution.

The Complex Exponential Polyreference limits its global solution to the modal parameters, $\omega_r$, $\zeta_r$ and $\underline{m}_r$. However, the Orthogonal Polyreference actually provides a global solution for the mode shapes as well.

# CHAPTER IX

# NYQUIST CIRCLE MODE EXTRACTION

## 9.1 Circle Mode Shape Extraction Method

As was demonstrated in Figure 7-9, a modal FRF may be plotted as the imaginary part versus the real part. This is referred to as a Nyquist plot. The SDOF FRF displays as a circle with frequency increasing along a clockwise path. The Nyquist data do not fit the equation of a circle precisely. However, Appendix A proves the data is a good approximation to a circle within the structural bandwidth. The plot of Figure 9-1 indicates aspects of exploiting the circle relationships for extraction of shapes, resonance frequency and damping factor. We focus first on circle curve fitting for mode shape extraction.

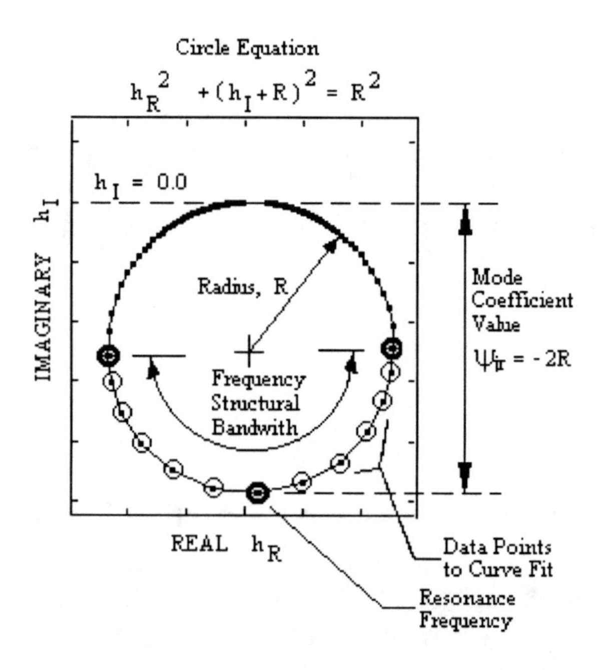


**Figure 9-1 FRF Nyquist Plot. Plotting the imaginary part versus the real part approximates a circle. The diameter is equal to the mode coefficient value for that location on the structure. A least squares curve fit could be performed using data points within the structural bandwidth. Resonance occurs at the maximum data point spacing, and the damping factor is a function of circumferential data point sweep rate.**

If the vertical axis of the Figure 9-1 graph is the FRF imaginary part, $h_I$, and the horizontal axis is the real part, $h_R$, then the equation of the circle with radius, R is

$$h_r^{\,2} + (h_I + R)^2 = R^2 \tag{9-1}$$

Typically, an equation like (9-1) will be curve fit to selected FRF data points that fall within the structural bandwidth (recall Figure 6-16 and equation 6-91). Once the equation has been fit to the data points, then the mode coefficient may be obtained directly from the radius, R. For a curve fit of FRF, $h_{jk}(\omega)$, the mode coefficient is $\Psi_{jr} = 2R$. It is also possible to rescale the FRF data so that the circle fit would allow one to solve for the damping ratio, $\zeta_r$. Then, $R = 1/(4\zeta_r)$.

Figure 9-2 is a Nyquist plot of an analytically computed FRF having four resonance frequencies.

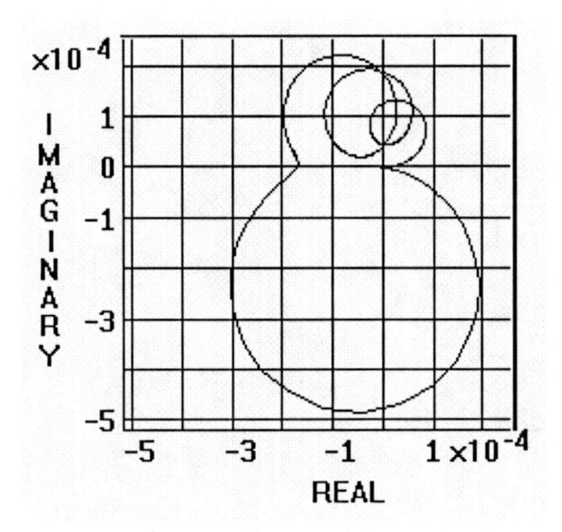


**Figure 9-2 A Nyquist plot of an analytically generated FRF having four resonance frequencies. The modes are fairly well separated in frequency so that four distinct circular patterns are evident. Each circular neighborhood may be fit using the equation of a circle.**

The FRF plot is characterized by four circular patterns, which correspond to the four modes. It is recognized that modal parameters for each of the modes may be obtained by fitting the SDOF FRF circle equation to each of the circles in the Nyquist plot. Notice how the Nyquist plot transitions from one circle to the next. The transition data points between one circle and the next are not useful for the circle fits. Only the data points conforming to a consistent circular pattern are used in the circle fits. Good circle fit software provides a user interface with interactive graphics allowing easy selection of data points to be used in the circle fit.

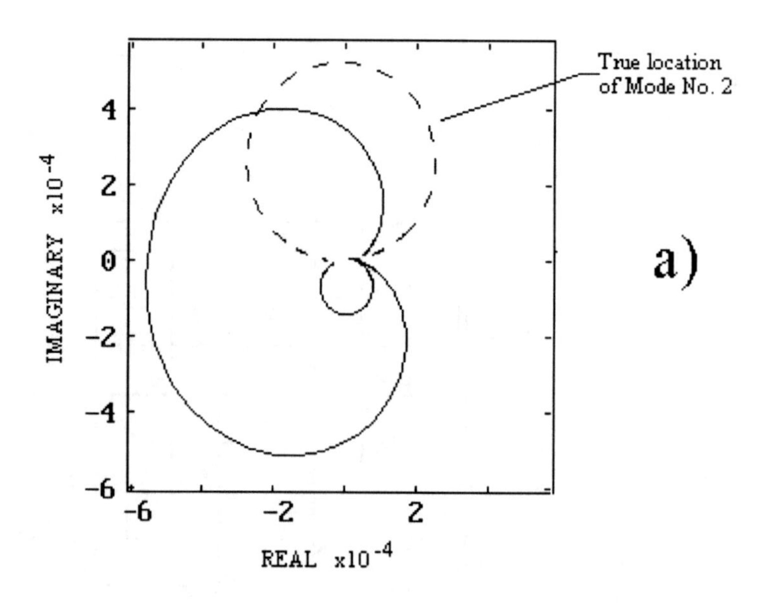


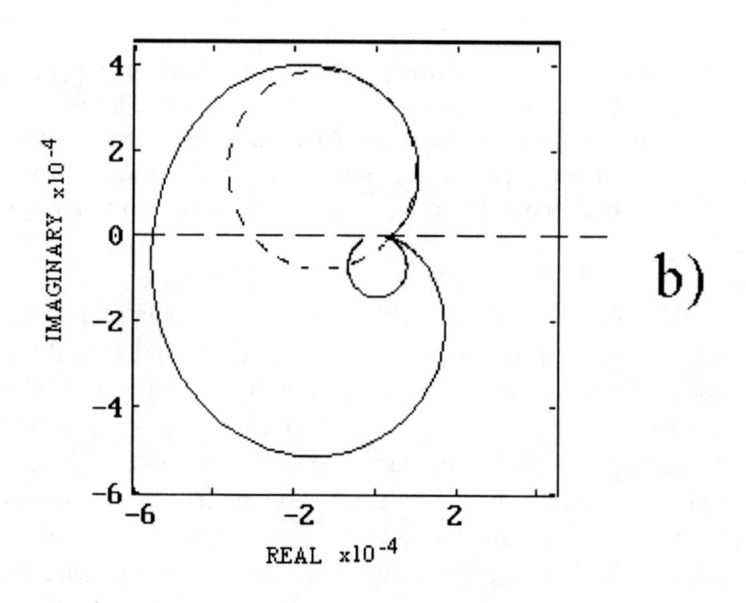

**Figure 9-3 Nyquist plot of FRF with three modes. The first two modes are closely spaced in frequency and the third mode is well separated. The dashed circle in the upper plot a) indicates the position of the true mode #2 Nyquist plot without effects of mode #1. The lower plot b) shows the position of the mode #2 Nyquist plot under the influence of mode #1. This is the same FRF as plotted in Figure 8-3 and sets up the strategy for extracting mode coefficients using the circle fit.**

Figure 9-2 illustrates the ease of associating modes with circles exhibited by a FRF Nyquist plot. That is because those modes are well separated in frequency. Figures 9-3 and 9-4 illustrate the process of extracting modal parameters of a given mode when resonance frequencies are closely spaced. The FRF in this case is the same as that of Figure 8-3, where we demonstrated the failure of the direct extraction method when resonances are close to each other.

Figure 9-3, upper plot a), shows the position of the circle representing mode number two, where mode two has been separated out of the FRF and is independent of and not influenced by mode one. The lower plot b) shows the position of the mode two circle under the influence of mode one.

Figure 9-4 shows the selection of data points to be included in the curve fit for mode number two.

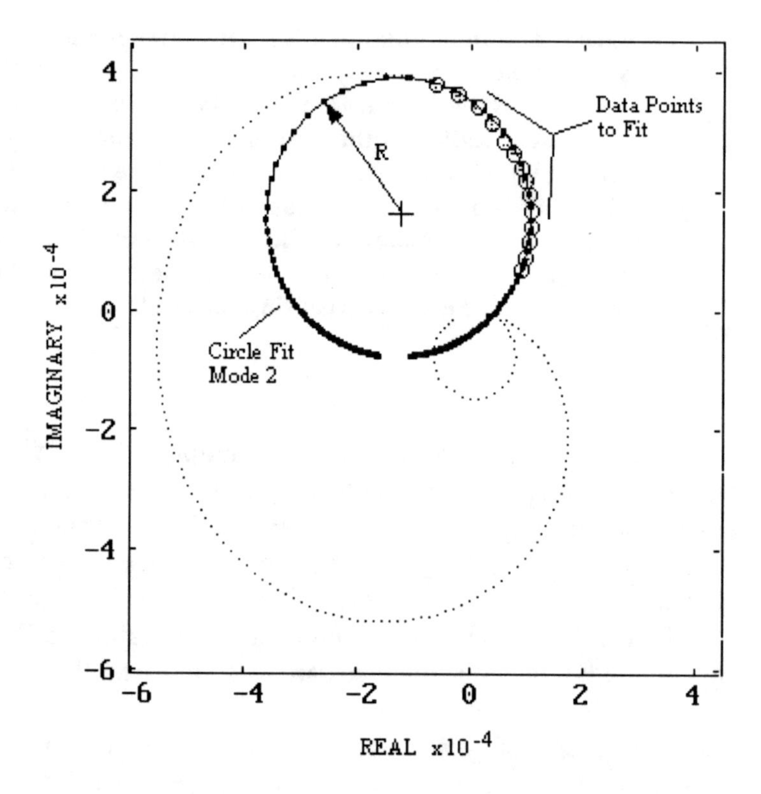


**Figure 9-4 Setting up the strategy for a circle curve fit of mode number two. Data points in the transition between one circle and the next must be avoided in the curve fit.**

## 9.2 Least Squares Solution for Mode Shapes

The FRFs of Figures 9-1 through 9-4 were generated mathematically as noise free idealized functions. The same FRF is made more realistic by adding random noise to the data and is displayed as a scatter plot in Figure 9-5. The best-fit position for the mode number two circle is shown, based on a least squares circle fit.

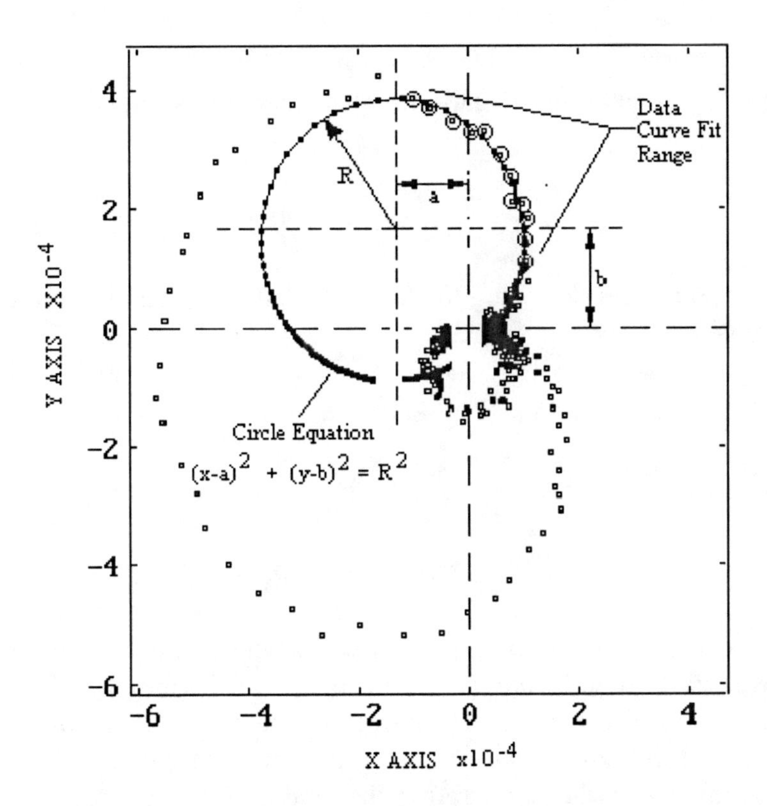


**Figure 9-5 Best fit for mode number two Nyquist circle. The data curve fit range is indicated on the graph. The form of the circle equation is indicated along with the definition of the circle parameters, a, b and R.**

It is seen that the circle matches the FRF data over a limited frequency range. The data curve fit range is indicated on the Figure 9-5 graph. The FRF real and imaginary axes are referenced as the variables, x and y. The circle equation parameters to be discovered by the circle fit process are a, b and R. The mode number two Nyquist circle is offset from the origin of the x-y coordinate system by an amount, a, in the x direction and b in the y direction. The radius of the mode number two circle is R. The circle equation is then

$$(x-a)^2+(y-b)^2=R^2 \tag{9-1}$$

Carrying out the multiplication of the squared binomials,

$$x^2+y^2-2ax-2by+a^2+b^2-R^2=0 \tag{9-2}$$

Regrouping and arranging to isolate the squared variables on the right hand side,

$$2ax+2by+(R^2-a^2-b^2)=x^2+y^2 \tag{9-3}$$

Introduce a change of variables for more concise notation.

$$Ax+By+C=x^2+y^2 \tag{9-4}$$

A set of n data points are selected for the circle fit. These points correspond to the data curve fit range of Figure 9-5. The circle equation (9-4) can be written n times, once for each data point within the curve fit range. This is exactly the same procedure followed with equation (8-9) when illustrating the straight-line example. Again, we write these equations in matrix form:

$$\begin{bmatrix} x_1 & y_1 & 1 \\ x_2 & y_2 & 1 \\ x_3 & y_3 & 1 \\ x_4 & y_4 & 1 \\ \vdots & \vdots & \vdots \\ x_n & y_n & 1 \end{bmatrix} \begin{Bmatrix} A \\ B \\ C \end{Bmatrix} = \begin{Bmatrix} x_1^2+y_1^2 \\ x_2^2+y_2^2 \\ x_3^2+y_3^2 \\ x_4^2+y_4^2 \\ \vdots \\ x_n^2+y_n^2 \end{Bmatrix} \tag{9-5}$$

The circle parameters, A, B and C are solved as before using the pseudo-inverse scheme.

$$\begin{Bmatrix} A \\ B \\ C \end{Bmatrix} = \left[ \begin{bmatrix} x_1 & y_1 & 1 \\ x_2 & y_2 & 1 \\ x_3 & y_3 & 1 \\ x_4 & y_4 & 1 \\ \vdots & \vdots & \vdots \\ x_n & y_n & 1 \end{bmatrix}^T \begin{bmatrix} x_1 & y_1 & 1 \\ x_2 & y_2 & 1 \\ x_3 & y_3 & 1 \\ x_4 & y_4 & 1 \\ \vdots & \vdots & \vdots \\ x_n & y_n & 1 \end{bmatrix} \right]^{-1} \begin{bmatrix} x_1 & y_1 & 1 \\ x_2 & y_2 & 1 \\ x_3 & y_3 & 1 \\ x_4 & y_4 & 1 \\ \vdots & \vdots & \vdots \\ x_n & y_n & 1 \end{bmatrix}^T \begin{Bmatrix} x_1^2 + y_1^2 \\ x_2^2 + y_2^2 \\ x_3^2 + y_3^2 \\ x_4^2 + y_4^2 \\ \vdots \\ x_n^2 + y_n^2 \end{Bmatrix}$$

(9-6)

Inserting the x and y values into the equation (9-6) for all FRF data points within the curve fit range reduces the problem to basic matrix operations. These operations are performed very rapidly on a computer.

After computing the values for A, B and C, the original circle parameters are

$$a = \frac{A}{2} \quad (9\text{-}7)$$

$$b = \frac{B}{2} \quad (9\text{-}8)$$

$$R = \sqrt{a^2 + b^2 + C} \quad (9\text{-}9)$$

Assuming this result has just been obtained for mode number two of some FRF, $h_{jk}(\omega)$,
then equation (9-9) also gives us the value of the mode coefficient, $\psi_{j2.}$ See Figure 9-1 and recall choice of scaling for $\psi_{jr}$ in equation 7-35.

$$\psi_{j2} = 2R = 2\sqrt{a^2 + b^2 + C} \tag{9-10}$$

All mode coefficients for mode number two may be obtained by curve fitting all of the FRFs over that same mode two frequency range.

Finally, all FRFs would be curve fit again, using a new frequency range corresponding to a different resonance frequency and mode shape. This process is repeated until all FRFs have been curve fit for every resonance frequency and mode shape of interest.

## 9.3 Circle Resonance Frequency Identification

One method of identifying a FRF resonance frequency is to take advantage of the unique spacing of fixed frequency increment data points on the Nyquist circle. Looking again at Figure 9-1, the data points have greater circumferential path separation the closer they are to the resonance frequency. Appendix B is devoted to proving that the data point of resonance is located, with good approximation, at the point of maximum circumferential path sweep rate with respect to change in frequency. This condition should hold fairly well even with a group of moderately close resonance frequencies.

This suggests that a resonance frequency may be identified using this sweep rate property as an improvement over the peak extraction method. The implementation of the method involves a systematic search of FRF data in the neighborhood of the resonance frequency. Computer software supporting this method provides a user interface with interactive graphics options allowing easy selection of the relevant frequency search range. Often, the same frequency range used for the mode shape circle fit is used for the resonance search. The resonance frequency may be identified using one strategically selected FRF, or it may be estimated from the average of a select FRF group.

The software algorithm would typically build a list of circular path distances from data point to data point. Then, the list is searched for the maximum distance. The straight-line distances are used and are computed using the Pythagorean distance formula. The distance from the data point, p-1, to the data point, p, is

$$d_p = \sqrt{\left[h_{Real}(\omega_p) - h_{Real}(\omega_{p-1})\right]^2 + \left[h_{Imag}(\omega_p) - h_{Imag}(\omega_{p-1})\right]^2}$$

(9-11)

We consider the data point index, p, to increment with the succession of points on the Nyquist curve moving clockwise. This same index, p, also provides the indexing of frequency values, $\omega_1$, $\omega_2$, $\omega_3$, ..., $\omega_p$. The frequency values are assumed to be separated by a fixed frequency increment, $\Delta\omega$, so that any frequency value within an FRF may be found as

$$\omega_p = p \bullet \Delta\omega, \quad p = 0,1,2,3,... \qquad (9\text{-}12)$$

Thus, if the maximum distance found is $d_m$, then the resonance frequency is identified as $\omega_m$.

Sometimes, FRF resonances have very light damping and are computed using relatively short data block time periods (total acquisition time for one block of time data). Consequently, there are large values of $\Delta\omega$ and long circumferential path lengths, resulting in only two or three data points within the structural bandwidth. This has led to various attempts at improving the estimate of resonance frequency using FRF interpolation methods. This increases the frequency resolution and adds more data points to the Nyquist circle.

Another scheme for increasing frequency resolution is to inverse Fourier Transform the FRF, getting the data back into the time domain. This yields the impulse response function. Here the computer program would add more data points to the time data block, setting a value of zero for all of the added points. Or else, add on a copy of the same impulse response function, attenuating it appropriately so the data continues from the same amplitude level corresponding to the end of the original data block. A smoothing function may be used to avoid a discontinuity between the end of the original data block and the start of the added-on data. The new time data block is then Fourier Transformed back into the frequency domain where the $\Delta\omega$ frequency increment is now reduced (by whatever factor the time domain was increased). Note that the frequency resolution, $\Delta\omega$, of the Fourier Transform of a time data block is

$$\Delta\omega = \frac{1}{2\pi T} \tag{9-13}$$

where T is the time period (duration) of the data block.

Additional data smoothing tricks of the trade may be employed to reduce the artifact that accompanies this Fourier Transform strategy. However, one would probably resort to other resonance identification methods before exerting too much effort in overcoming any problems with the circle method.

The frequency resolution problem should really be addressed on the front end of a modal test project. Data acquisition parameters should be planned to assure a sufficient number of data points fall within the structural bandwidth. Sometimes a zoom transform is used at higher frequencies.

### 9.4 Circle Damping Factor Identification

Once the resonance frequency is known, the Nyquist circle data angular sweep rate can be used to estimate the damping factor. Equation (B-8) of Appendix B provides the angular sweep rate formula,

$$\frac{d\theta}{d\omega} = -\frac{2\zeta_r\omega_r(\omega_r^{\ 2}+\omega^2)}{(\omega_r^{\ 2}-\omega^2)+4\zeta_r^{\ 2}\omega_r^{\ 2}\omega^2} \qquad (9\text{-}14)$$

Evaluating the angular sweep rate at the resonance frequency, $\omega = \omega_r$,

$$\left.\frac{d\theta}{d\omega}\right|_{\omega=\omega_r} = \frac{1}{\zeta_r\omega_r} \qquad (9\text{-}15)$$

And $\zeta_r$ is computed as

$$\zeta_r = \frac{1}{\omega_r} \bullet \frac{\Delta\omega}{\Delta\theta} \qquad (9\text{-}16)$$

This formula is valid for $\Delta\omega$ and $\Delta\theta$ computed using a pair of data points on the Nyquist circle in the neighborhood of the resonance frequency. Figure 9-6 illustrates the computation. Due to random errors in the measurement process there are often slight fluctuations in the data point positions relative to the true Nyquist circle as was illustrated in Figure 9-5. It is best to obtain an averaged estimate of $\zeta_r$ by computing the formula of (9-16) using as large a $\Delta\theta$ range as practical. Even though the Equation (9-15) indicates the formula was derived specifically for the resonance frequency point on the Nyquist circle, it has been found that good approximations are obtained using a $\Delta\theta$ as large as fifteen to twenty degrees.

The process illustrated in Figure 9-6 indicates an estimate over the range of two FRF data points labeled as p-2 and p. The $\Delta\theta$ value would be computed using the trigonometric arctangent function for each of the points:

$$\Delta\theta = \tan^{-1}\left(\frac{h_I(\omega_p)}{h_R(\omega_p)}\right) - \tan^{-1}\left(\frac{h_I(\omega_{p-2})}{h_R(\omega_{p-2})}\right) \qquad (9\text{-}17)$$

The $\Delta\omega$ value is of course just

$$\Delta\omega = \omega_p - \omega_{p-2} \tag{9-18}$$

The number of data points to span for use in the $\Delta\theta$ and $\Delta\omega$ computations depends on the density of points along the Nyquist circle in the neighborhood of resonance.

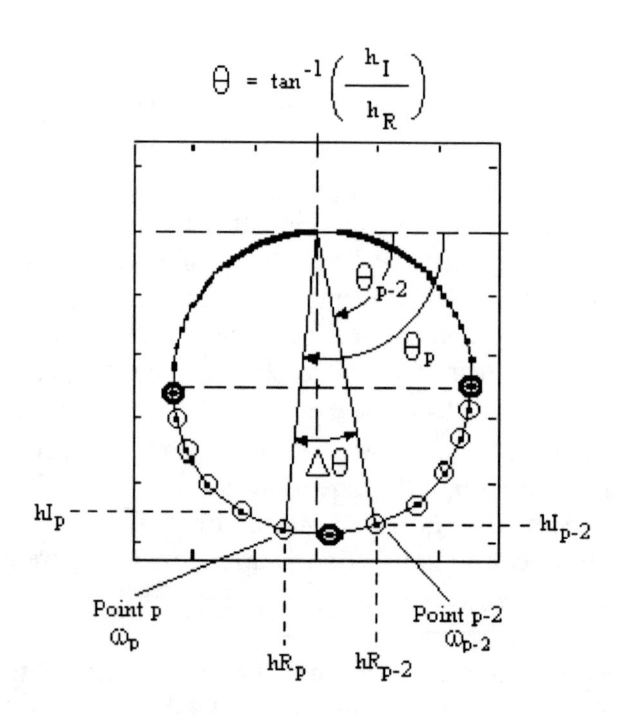


**Figure 9-6 The measurement of Nyquist FRF circle parameters needed for computing the damping factor, $\zeta_r$, is illustrated (see Eq 9-16). The $\Delta\theta$ uses arctangent values for two sets of real and imaginary FRF values corresponding to points p-2 and p in the sketch. $\Delta\omega = \omega_p - \omega_{p-2}$.**

## 9.5 Improved Circle Damping Estimate

The trouble with our damping estimate up to this point is that one of the assumptions of Equation (9-14) may not hold up well in some instances. The angular sweep rate formula of Equation (9-14) and the resulting damping formula of Equation (9-16) assume that the data falls on a Nyquist circle that is in a standard isolated mode position. However, it was seen in Figure 9-3 that when you have a MDOF system a Nyquist circle can be shifted in the Argand plane under the influence of neighboring modes. Figure 9-7 illustrates the problem with computing the $\Delta\theta$ angle of Figure 9-6 when the circle has been shifted under the influence of neighboring modes. Figure 9-7 (A) is the offset circle, and the angle indicated as theta prime, θ', is the angle that would be computed using the arctangent formulas of Equation (9-17).

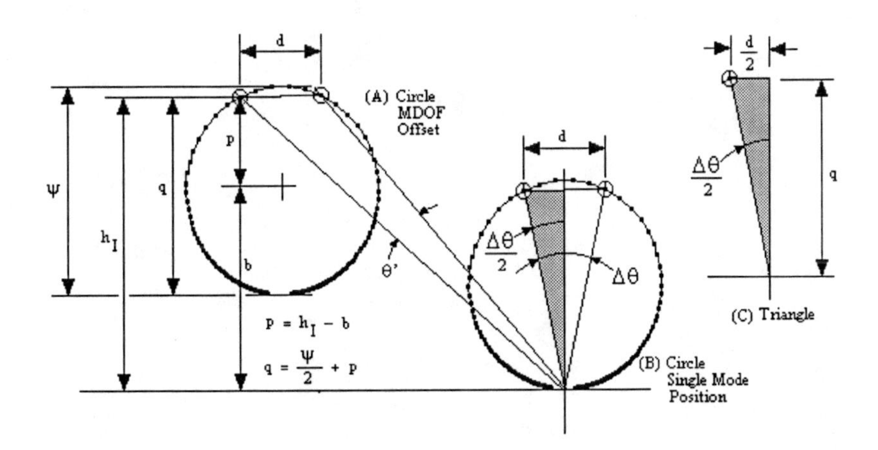


**Figure 9-7 An improved method of computing Δθ for estimating the damping fraction using Equation (9-16) is illustrated. This applies when the Nyquist circle has been offset as in (A) under the influence of neighboring modes. Otherwise, the angle, θ', would be computed if Equation (9-17) is used. The isolated mode position of (B) is the underlying assumption of Equation (9-14). The triangle in (C) is used to compute Δθ.**

Equation (9-16) would be computed with better accuracy using the angle, $\Delta\theta$, shown in Figure 9-7 (B). The shaded triangle of Figure 9-7 (B) is replicated as Figure 9-7 (C) to clarify the more accurate computation of $\Delta\theta$ using the trigonometric arctangent function.

$$\Delta\theta = 2\tan^{-1}\left(\frac{d}{2q}\right) \tag{9-19}$$

The distance, d, between the two data points is computed using the Pythagorean distance formula as before in Equation (9-11). The length of the side of the triangle, q, is computed using results of the earlier circle curve fit for $\psi$ and b. The value, $\psi$, is the diameter of the circle, and b is the vertical offset of the center of the circle from the Argand plane origin. See Equations (9-8) and (9-10). The Figure 9.7 (A) construction provides the computations for p and q:

$$p = h_{Imag} - b \tag{9-20}$$

$$q = \frac{\psi}{2} + p \tag{9-21}$$

# APPENDIX A

# FRF CIRCLE EQUATION

## A.1 FRF Nyquist Plot Observations

As was demonstrated in Figure 6-9 and again in Figure 9-1, a modal FRF representing a SDOF system may be plotted in the form of the imaginary part versus the real part. This type of plot is referred to as a Nyquist plot (also, Argand plot). The plot has the appearance of a circle with frequency increasing along the path of the circle going clockwise. The Nyquist plot of Figure 9-1 is plotted again here as Figure A.1.

It is the purpose of this Appendix to show that the FRF Nyquist plot of a SDOF system with viscous damping approximates a circle. It will be shown that the approximation is very good within the frequency range of the structural bandwidth when the structure is lightly damped.

A FRF Nyquist plot of a SDOF system with structural damping (hysteretic damping) accurately displays as a circle. However, most curve-fit methods and experimental modal models discussed in this text will be represented using viscous damping.

Some of the earlier SDOF relationships were actually based on the assumption of a circular FRF Nyquist plot. For example the upper frequency boundary for the structural bandwidth, $\omega_U$, has been given as (from Equation 6-89):

$$\omega_U \cong \omega_r (1+\zeta_r) \tag{A-1}$$

where $\omega_r$ (Greek lower case omega) is the resonance frequency (radians/sec) and $\zeta_r$ (Greek lower case zeta) is the viscous damping fraction.

This also led to the relationship between damping fraction, resonance frequency and structural bandwidth:

$$\zeta_r \cong \frac{SB}{2\omega_r} \qquad \text{(A-2)}$$

And clearly, if the Nyquist plot of a SDOF system is a circle, the absolute magnitudes of the maximum and minimum real FRF values are one-half of the magnitude of the imaginary FRF extreme value.

The Nyquist plot of Figure A.1 was generated directly from the SDOF FRF formula. But, in spite of the qualitative appearance of a circle, it should not be automatically assumed that the data accurately define a circle. In view of the significance of the many results that will be based on the Nyquist circle assumption, it is not being overly cautious to provide mathematical support for this assumption.

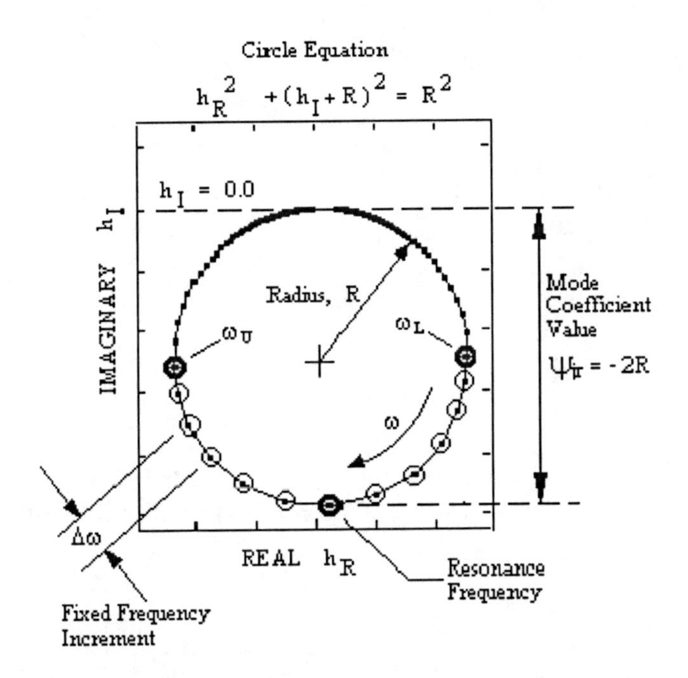


**Figure A.1 FRF Nyquist Plot. Plotting the imaginary part versus the real part approximates a circle. The approximation is particularly good within the structural bandwidth, i.e., in the frequency range between $\omega_L$ and $\omega_U$. Frequency increases clockwise along the circular path.**

## A.2 Mathematical Development Of The Nyquist Circle

We now show that the Nyquist FRF data points approximately fit the equation of a circle. Starting with the FRF formula for just one mode, r, extracted from the r.h.s. summation of Equation (7-23),

$$h_{jk}^{r}(\omega) = \frac{\psi_{jr}\psi_{kr}}{\omega_r^{2} m_r}\left[\frac{(1-\beta_r^{2}) - 2i\zeta_r\beta_r}{(1-\beta_r^{2})^2 + 4\zeta_r^{2}\beta_r^{2}}\right] \qquad \text{(A-3)}$$

The constants in front of the r.h.s. bracket will be lumped into a single scaling constant, $A_{jk}^{r}$.

$$h_{jk}^{r}(\omega) = A_{jk}^{r}\left[\frac{(1-\beta_r^{2}) - 2i\zeta_r\beta_r}{(1-\beta_r^{2})^2 + 4\zeta_r^{2}\beta_r^{2}}\right] \qquad \text{(A-4)}$$

For purposes of developing the Nyquist circle formula, it will be convenient to rescale Equation (A-4) by dividing through by $A_{jk}^{r}$.

$$\frac{h_{jk}^{r}(\omega)}{A_{jk}^{r}} = \frac{(1-\beta_r^{2}) - 2i\zeta_r\beta_r}{(1-\beta_r^{2})^2 + 4\zeta_r^{2}\beta_r^{2}} \qquad \text{(A-5)}$$

Showing that the Nyquist plot of Equation (A-5) approximately fits the equation of a circle is the equivalent of demonstrating the circle property for the FRF Nyquist plot of Equation (A-3). Rescaling simply changes the size of the circle. An inspection of (A-5) at the resonance frequency, $\omega_r$, where $\beta_r = 1$ shows that the diameter of the rescaled circle is

$$\frac{h_{jk}^{r}(\omega)}{A_{jk}^{r}} = \frac{1}{2\zeta_r} \qquad \text{(A-6)}$$

This implies that the Nyquist plot for Equation (A-5) fits a circle equation of the form

$$X^2 + \left(Y + \frac{1}{4\zeta_r}\right)^2 = \left(\frac{1}{4\zeta_r}\right)^2 \qquad \text{(A-7)}$$

where the circle radius is $1/(4\zeta_r)$ and the center of the circle is offset in the negative Y direction by the distance of the circle radius.

Assuming a Nyquist circle, the relationship between the structural bandwidth upper frequency bound, $\omega_U$, and the damping factor, $\zeta_r$, and resonance frequency, $\omega_r$, are the same as expressed in Equations (A-1) and (A-2). The derivation of these relations was initiated by setting the FRF real and imaginary parts equal in Equation (6-85), then solving for $\omega_U$ and $\omega_L$. It is seen that setting the real and imaginary parts equal in Equation (A-5) above yields exactly the same starting point as Equation (6-85).

Now, it will not be possible to prove Equation (A-7), because the Nyquist plot only approximates a circle equation. Instead, it will be shown that the following approximation to a circle equation holds within the structural bandwidth:

$$X^2 + \left(Y + \frac{1}{4\zeta_r\beta_r}\right)^2 = \left(\frac{1}{4\zeta_r\beta_r}\right)^2 \tag{A-8}$$

Maintaining this approximation within the frequency range of the structural bandwidth means that the worst error would occur at the upper or lower edge of the structural bandwidth. Remember that at the upper frequency bound, $\omega_U$, of the structural bandwidth, the value of $\beta_r$ is very close to 1.0. That is, repeating Equation (A-1)

$$\omega_U \cong \omega_r(1+\zeta_r) \tag{A-9}$$

Or, at $\beta_r = \omega_U/\omega_r$,

$$\beta_r \cong 1+\zeta_r \tag{A-10}$$

The situation is of course the same at the lower bound of the structural bandwidth, $\omega_L$. For lightly damped structures, $\zeta_r$ would be the order of

0.01 and $\beta_r$ would be approximately equal to 1.0, within about 1% error, depending on the value of $\zeta_r$.

To show that the Equation (A-8) approximation to a circle is valid, the real and imaginary FRF functions will be substituted for X and Y. First, the Y term of the circle equation is manipulated.

$$Y+\frac{1}{4\zeta_r\beta_r}=\frac{-2\zeta_r\beta_r}{\left(1-\beta_r^{\ 2}\right)^2+4\zeta_r^{\ 2}\beta_r^{\ 2}}+\frac{1}{4\zeta_r\beta_r} \qquad \text{(A-11)}$$

$$Y+\frac{1}{4\zeta_r\beta_r}=\frac{(-2\zeta_r\beta_r)(4\zeta_r\beta_r)+\left[\left(1-\beta_r^{\ 2}\right)^2+4\zeta_r^{\ 2}\beta_r^{\ 2}\right]}{4\zeta_r\beta_r\left[\left(1-\beta_r^{\ 2}\right)^2+4\zeta_r^{\ 2}\beta_r^{\ 2}\right]} \qquad \text{(A-12)}$$

$$Y+\frac{1}{4\zeta_r\beta_r}=\frac{\left(1-\beta_r^{\ 2}\right)^2-4\zeta_r^{\ 2}\beta_r^{\ 2}}{4\zeta_r\beta_r\left[\left(1-\beta_r^{\ 2}\right)^2+4\zeta_r^{\ 2}\beta_r^{\ 2}\right]} \qquad \text{(A-13)}$$

Next, square both sides of (A-13) to obtain an expression for the squared term in (A-8).

$$\left(Y+\frac{1}{4\zeta_r}\right)^2=\frac{\left(1-\beta_r^{\ 2}\right)^4-8\zeta_r^{\ 2}\beta_r^{\ 2}\left(1-\beta_r^{\ 2}\right)^2+16\zeta_r^{\ 4}\beta_r^{\ 4}}{\left\{4\zeta_r\beta_r\left[\left(1-\beta_r^{\ 2}\right)^2+4\zeta_r^{\ 2}\beta_r^{\ 2}\right]\right\}^2} \qquad \text{(A-14)}$$

Working for a moment with the X term,

$$X = \frac{1-\beta_r^{\ 2}}{\left(1-\beta_r^{\ 2}\right)^2 + 4\zeta_r^{\ 2}\beta_r^{\ 2}} \tag{A-15}$$

Squaring both sides of (A-15) and multiplying top and bottom of r.h.s. by the factor needed to obtain a common denominator with Equation (A-14),

$$X^2 = \frac{16\zeta_r^{\ 2}\beta_r^{\ 2}\left(1-\beta_r^{\ 2}\right)^2}{\left\{4\zeta_r\beta_r\left[\left(1-\beta_r^{\ 2}\right)^2 + 4\zeta_r^{\ 2}\beta_r^{\ 2}\right]\right\}^2} \tag{A-16}$$

Now, substitute these results, the right hand sides of equations (A-16) and (A-14) in place of the X and Y terms on the left hand side of Equation (A-8), taking advantage of the common denominators.

$$\frac{16\zeta_r^{\ 2}\beta_r^{\ 2}\left(1-\beta_r^{\ 2}\right)^2 + \left(1-\beta_r^{\ 2}\right)^4 - 8\zeta_r^{\ 2}\beta_r^{\ 2}\left(1-\beta_r^{\ 2}\right)^2 + 16\zeta_r^{\ 4}\beta_r^{\ 4}}{\left\{4\zeta_r\beta_r\left[\left(1-\beta_r^{\ 2}\right)^2 + 4\zeta_r^{\ 2}\beta_r^{\ 2}\right]\right\}^2} = \left(\frac{1}{4\zeta_r\beta_r}\right)^2 \tag{A-17}$$

Collecting terms in (A-17),

$$\frac{\left(1-\beta_r^{\,2}\right)^4+8\zeta_r^{\,2}\beta_r^{\,2}\left(1-\beta_r^{\,2}\right)^2+\left(4\zeta_r^{\,2}\beta_r^{\,2}\right)^2}{\left(4\zeta_r\beta_r\right)^2\left[\left(1-\beta_r^{\,2}\right)^2+4\zeta_r^{\,2}\beta_r^{\,2}\right]^2}=\left(\frac{1}{4\zeta_r\beta_r}\right)^2 \qquad \text{(A-18)}$$

Notice that the left hand side numerator may be factored as a perfect square expression.

$$\frac{\left[\left(1-\beta_r^{\,2}\right)^2+4\zeta_r^{\,2}\beta_r^{\,2}\right]^2}{\left(4\zeta_r\beta_r\right)^2\left[\left(1-\beta_r^{\,2}\right)^2+4\zeta_r^{\,2}\beta_r^{\,2}\right]^2}=\left(\frac{1}{4\zeta_r\beta_r}\right)^2 \qquad \text{(A-19)}$$

Canceling out the common expressions in numerator and denominator of (A-19),

$$\left(\frac{1}{4\zeta_r\beta_r}\right)^2=\left(\frac{1}{4\zeta_r\beta_r}\right)^2 \qquad \text{(A-20)}$$

This completes the proof that the FRF Nyquist plot for a SDOF system approximates a circular plot obeying Equation (A-8) within the structural bandwidth.

# APPENDIX B

# NYQUIST PLOT RESONANCE LOCATION

## B.1 Nyquist Plot Frequency Spacing

As was demonstrated in Figure 6-9 and again in Figure 9-1, a modal FRF representing a SDOF system may be plotted in the form of the imaginary part versus the real part. This type of plot is referred to as a Nyquist plot (also, Argand plot). It displays as a circle with frequency increasing along the path of the circle going clockwise. The Nyquist plot of Figure 9-1 is plotted again here as Figure B.1. The upper and lower structural bandwidth frequencies, $\omega_U$ and $\omega_L$, are indicated (refer to Equations 6-89 and 6-90).

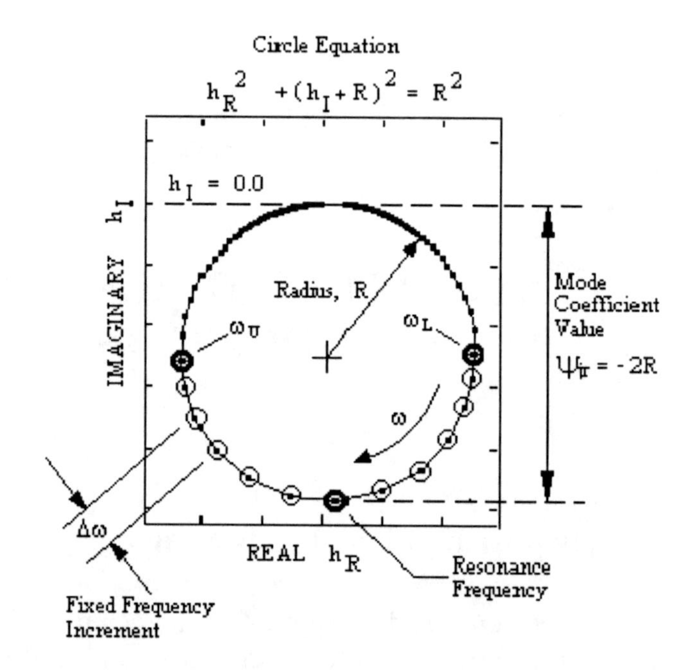


**Figure B.1 FRF Nyquist Plot. Plotting the imaginary part versus the real part approximates a circle. The resonance frequency is located at the point on the circle where the arc path spacing is the greatest for fixed frequency increments.**

Since the data points are plotted around the circle with a constant delta frequency from point to point, it is seen by inspecting Figure B.1 that the resonance frequency occurs at the bottom of the circle where there is maximum circular path distance between data points. The location of the resonance is a trivial issue for the case that the mode of interest is manifest as an isolated circle. However, when an FRF with multiple closely spaced modes are involved, the position of maximum frequency increment spacing may be shifted to the right or left with respect to the bottom of the circle.

While it seems obvious from the Figure B.1 plot that resonance occurs at the maximum spacing, that does not constitute proof. The purpose of this Appendix is to show analytically that the resonance frequency occurs at the position on the circle at which the ratio of delta path length to delta frequency is a maximum.

Rather than use the circumferential path length directly, the angle theta, $\theta$, will be used. Theta will be the angle from the origin in the FRF Argand plane to a FRF data point on the Nyquist circle. The angle, $\theta_i$, is shown for the $i^{th}$ data point in Figure B.2.

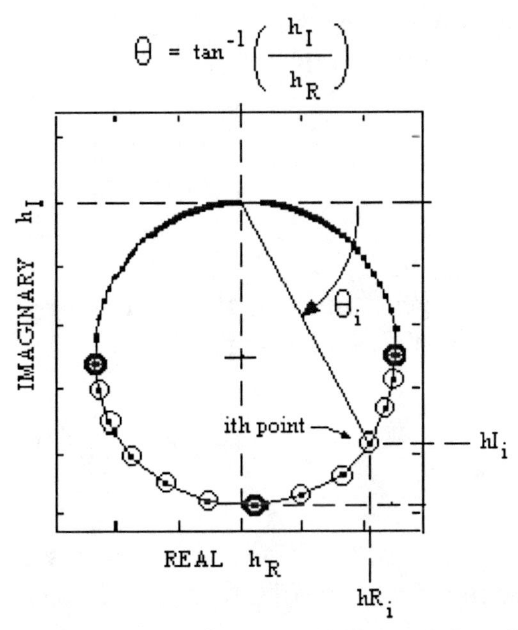


**Figure B.2 Angle theta, $\theta_i$, is is illustrated for the ith Nyquist plot FRF data point. The resonance frequency will be shown to occur at the position of maximum change of angle with respect to delta frequency.**

The angle theta relates to the FRF real and imaginary values through the trigonometric arctangent function.

$$\theta = \tan^{-1}\left(\frac{h_I(\omega)}{h_R(\omega)}\right) \tag{B-1}$$

The maximum sweep rate position for the resonance frequency will be proven by setting the second derivative of theta with respect to frequency equal to zero and solving for the frequency, ω. The frequency, ω, will be shown to be equal to the resonance frequency, $\omega_r$, at that point.

Starting with the FRF formula for just one mode, r, extracted from the r.h.s. summation of Equation (7-23),

$$h^r_{jk}(\omega) = \frac{\psi_{jr}\psi_{kr}}{\omega_r^{\,2} m_r}\left[\frac{(1-\beta_r^{\,2}) - 2i\zeta_r\beta_r}{(1-\beta_r^{\,2})^2 + 4\zeta_r^{\,2}\beta_r^{\,2}}\right] \tag{B-2}$$

Substituting the real and imaginary parts of the Equation (B-2) FRF into the tangent function of Equation (B-1), we have after some algebraic manipulation using $\beta_r = \omega/\omega_r$.

$$\theta = \tan^{-1}\left(\frac{-2\zeta_r\omega_r\omega}{\omega_r^{\,2} - \omega^2}\right) \tag{B-3}$$

Let the tangent of Equation (B-3) be represented by the variable, u, so that the first derivative of θ with respect to ω is

$$\frac{d\theta}{d\omega} = \frac{1}{1+u^2} \bullet \frac{du}{d\omega} \tag{B-4}$$

Working for a moment with the derivative, du/dω,

$$u = -(2\zeta_r\omega_r\omega)(\omega_r^{\,2} - \omega^2)^{-1} \tag{B-5}$$

$$\frac{du}{d\omega} = \frac{-2\zeta_r\omega_r}{(\omega_r^{\ 2} - \omega^2)} - \frac{4\zeta_r\omega_r\omega^2}{(\omega_r^{\ 2} - \omega^2)^2} \tag{B-6}$$

Substituting (B-5) and (B-6) back into Equation (B-4),

$$\frac{d\theta}{d\omega} = -\frac{(\omega_r^{\ 2} - \omega^2)^2}{(\omega_r^{\ 2} - \omega^2)^2 + 4\zeta_r^{\ 2}\omega_r^{\ 2}\omega^2}\left[\frac{2\zeta_r\omega_r(\omega_r^{\ 2} - \omega^2) + 4\zeta_r\omega_r\omega^2}{(\omega_r^{\ 2} - \omega^2)^2}\right] \tag{B-7}$$

Dividing out common binomial squared expressions in numerator and denominator,

$$\frac{d\theta}{d\omega} = -\frac{2\zeta_r\omega_r(\omega_r^{\ 2} + \omega^2)}{(\omega_r^{\ 2} - \omega^2)^2 + 4\zeta_r^{\ 2}\omega_r^{\ 2}\omega^2} \tag{B-8}$$

Proceeding with the second derivative,

$$\frac{d^2\theta}{d\omega^2} = \frac{-4\zeta_r\omega_r\omega}{(\omega_r^{\ 2} - \omega^2)^2 + 4\zeta_r^{\ 2}\omega_r^{\ 2}\omega^2} + \frac{2\zeta_r\omega_r(\omega_r^{\ 2} + \omega^2)\left[-4\omega(\omega_r^{\ 2} - \omega^2) + 8\zeta_r^{\ 2}\omega_2^{\ 2}\omega\right]}{\left[(\omega_r^{\ 2} - \omega^2)^2 + 4\zeta_r^{\ 2}\omega_r^{\ 2}\omega^2\right]^2} \tag{B-9}$$

Now, the second derivative is set equal to zero. This is the condition for the maximum angular sweep rate. The frequency value satisfying this equation corresponds to the point on the Nyquist plot of maximum sweep rate. Having set (B-9) equal to zero, the squared denominator expression may be multiplied through. And performing further factoring, we have

$$4\zeta_r\omega_r\omega\left[(\omega_r^2-\omega^2)^2+4\zeta_r^2\omega_r^2\omega^2\right]-8\zeta_r\omega_r\omega(\omega_r^2+\omega^2)\left[(\omega_r^2-\omega^2)-2\zeta_r^2\omega_r^2\right]=0$$

(B-10)

Dividing out the common factor,

$$(\omega_r^2-\omega^2)^2+4\zeta_r^2\omega_r^2\omega^2-2(\omega_r^2+\omega^2)\left[(\omega_r^2-\omega^2)-2\zeta_r^2\omega_r^2\right]=0$$

(B-11)

We already know from the Appendix A endeavor that the Nyquist plot is only an approximation to a circle within the structural bandwidth. Likewise, the maximum sweep rate will not occur precisely at the resonance frequency. Rather than attempt an exact solution for $\omega$ at this point, an approximate solution will be found. Terms in equation (B-11) that include the square of the damping factor, $\zeta_r^2$, should be expected to be very small compared to all other terms ($\zeta_r^2$ should generally be of order $10^{-4}$). Setting all $\zeta_r^2$ terms equal to zero,

$$(\omega_r^2-\omega^2)^2-2(\omega_r^2+\omega^2)(\omega_r^2-\omega^2)=0 \quad \text{(B-12)}$$

$$(\omega_r^2-\omega^2)\left[(\omega_r^2-\omega^2)-2(\omega_r^2+\omega^2)\right]=0 \quad \text{(B-13)}$$

Expanding out into individual terms,

$$-\omega_r^4-3\omega_r^2\omega^2+\omega_r^2\omega^2+3\omega^4=0 \quad \text{(B-14)}$$

$$-\omega_r^4-2\omega_r^2\omega^2+3\omega^4=0 \quad \text{(B-15)}$$

By inspection it is seen that (B-15) has a solution for $\omega=\omega_r$. That is, if $\omega_r$ is substituted for $\omega$ the equation is satisfied within the desired approximation.

$$-\omega_r^4 - 2\omega_r^2\omega_r^2 + 3\omega_r^4 = 0 \tag{B-16}$$

This concludes the proof that the resonance frequency is located on the FRF Nyquist circle at the position of maximum angular sweep rate with respect to fixed frequency increments along the circular path. This further implies that the resonance occurs at the maximum circumferential path distance between fixed frequency increments.

Printed in the United States
103022LV00005B/124-162/A